Thinking Analytically

A GUIDE FOR MAKING DATA-DRIVEN DECISIONS

Jim Frost

Statistics By Jim Publishing

STATE COLLEGE, PENNSYLVANIA

U.S.A.

Published by: Statistics By Jim Publishing

State College, PA 16801

To contact the author, please email: jim@statisticsbyjim.com.

Visit the author's website at statisticsbyjim.com.

Ordering Information: Quantity sales. Special discounts are available on quantity purchases by educators. For details, contact the email address above.

Thinking Analytically: A Guide for Making Data-Driven Decisions / Jim Frost. —1st ed.

ISBN 979-8-9911935-0-4

Contents

Introduction

In a world inundated with data, the ability to analyze and understand information has never been more critical. Every day, we encounter a barrage of data and statistical claims in the media, advertising, and professional settings. Analytical thinking is both an art and a science, involving drawing meaningful conclusions from data.

"Thinking" is the first word in this book's title and that is its primary focus. I wrote this book to equip readers with the mindset and tools for thinking analytically. How do you think about problems, use information, and come to unbiased conclusions?

My book guides you in using data to draw valid conclusions and make informed decisions while avoiding common pitfalls. To accomplish this, you'll need to overcome challenges ranging from cognitive errors to analytical issues.

One thing this book *won't* do is get into the nitty-gritty details of various statistical analyses, such as when and how to use them and their assumptions—other books cover those topics extensively. While these details are crucial, this book focuses on broader issues affecting data and analysis in general. These include essential details impacting broad forms of assessment, even informal observations in the world.

They're the things you must always consider. How to *think* about problems that involve weighing information.

The world is complex, and understanding it through a limited set of numbers is challenging and fraught with pitfalls. One key analytical challenge is the human factor. Our brains are not naturally wired to process large amounts of data or accurately assess probabilities, leading to misguided conclusions and potential errors. In the worst cases, people with agendas can intentionally misuse data analysis for persuasion.

We live in the era of big data, where numbers represent a significant portion of our information. However, before this information can be beneficial, it must be collected, analyzed, interpreted, and mentally integrated—a process that educational systems have frequently neglected, producing analytical shortcomings. This book guides you in overcoming this deficiency, while the real world will provide ample opportunities to apply what you've learned.

I've written this book so it will be helpful for a wide variety of people interested in thinking analytically.

You don't need to be a practicing analyst to benefit from this book— just someone who wants to be an informed consumer of information and analysis. For instance, decision-makers who want to be more data-driven will find it helpful because I provide a mindset, technique, and examples for critiquing findings and things to watch out for.

I've also designed it for professionals who analyze data in various contexts, such as data analysts, big data specialists, machine learning, artificial intelligence, and those relying on data for decision-making in corporate and government settings. Additionally, it applies to scientists and researchers in academia, research institutions, and think tanks who use data to drive scientific discoveries and insights.

I wrote this book for several main reasons.

First, I found myself applying statistical thinking while out in the real world. That's not to say I walk around performing t-tests or regression analysis in my head. But there are statistical ways of thinking about the information you observe in everyday settings, such as how your observations can bias your mental framework or how natural variability and confounders distort your conclusions. Innate human capabilities often garble our views, whereas a carefully cultivated analytical outlook can help you think more clearly about such things in your daily life.

The same mindset helps me think critically about new discoveries and studies reported in the media. It also helps me evaluate arguments made by proponents of various viewpoints.

Second, in this big data era, more people are analyzing data and consuming the results than ever. While interest in analytical methods have increased, I've noticed that foundational principles are often neglected. Many analysts are adept at performing analyses but risk overlooking crucial concepts needed to analyze and interpret data effectively. Indeed, I see data analysts pay a ton of attention to algorithms and analyses, when to use them, and so on. But there's much less attention to underlying issues affecting their results, many of which occur before the analysis.

The thinking analytically mindset focuses on understanding the entire process behind real-world observations, information, and formalized data—all essential to avoid severe consequences from improper conclusions.

This book addresses key themes in thinking analytically so you can solve problems and make decisions effectively. In some cases, this process involves data analysis, but not always. I have structured the book to follow a five-level Thinking Analytically pyramid where each level builds on the next. Starting at the base and moving up, the levels are the following:

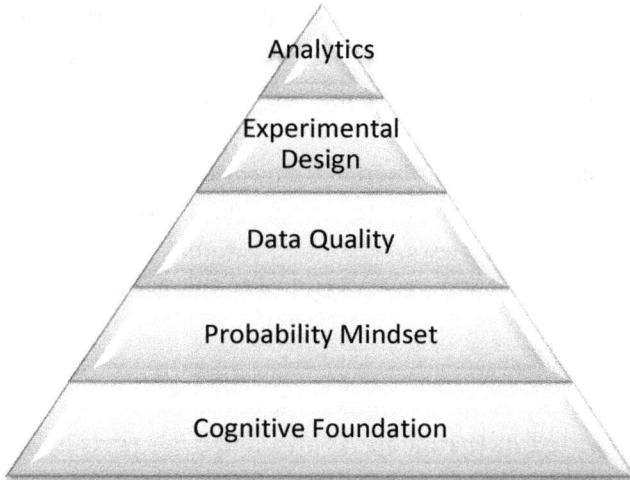

1. **Cognitive Foundation:** Understanding human cognitive limitations and how they influence our effectiveness in using and interpreting information.
2. **Probability Mindset:** The next level focuses on how well we intuitively understand probabilities. It covers common pitfalls and misconceptions in probability estimation and how to improve our probabilistic thinking. Are we good at estimating and using probabilities to understand the world?
3. **Data Quality:** Investigates the framework for ensuring data quality, including methods for collecting and measuring reliable data and understanding its limitations—all crucial for accurate analysis.
4. **Experimental Design:** Experimental designs determine what conclusions analysts can draw and how confident we can be that outside variables aren't distorting the results.
5. **Analytics:** We finally reach data analysis at the top of the pyramid. The validity of the results depends on the previous levels plus various methodological details common to many analyses.

While I focus generally on the thinking analytically mindset, I include several ongoing subthemes as well. These include the following:

Science and data analysis are hard because there are tons of details to get right. There are many ways to get the wrong answer and relatively few ways to get the right answer. Further, even when everything is correct, random chance can cause the analysis to produce incorrect results. That's why analysis requires so much care and attention. You'll see what it takes to perform good analyses.

Successful analytical thinking requires integrating the theories you are evaluating (i.e., the big ideas) with the nitty-gritty details. Theory and details must work together to have a chance at understanding the underlying reality. Throughout this book, you'll see many examples of where this integration failed and the resulting problems.

Using better approaches and methods will improve the quality of analytical results. However, that generally increases the time and cost to conduct a study. Sometimes, the best approaches are impossible. Understanding what to expect from data is crucial. Real-world data are more complicated than theory. For these reasons, you'll need to understand the limitations of less-than-ideal analyses because they are common.

Transparency and feedback about data and analytical methods are crucial for improving the results. Throughout this book, you'll see multiple ways science got something wrong. But it is a self-correcting field over time, and you'll see examples of that too.

Big data solves problems relating to sample size, but it tends to be more susceptible to factors that introduce bias. Unfortunately, massive datasets don't reduce bias.

Here are several other notes about this book.

I cover analytical thinking broadly, including your personal experiences in the world and in the context of data analysis. While there is a lot of overlap between these settings, the text ebbs and flows between them.

You'll notice that the data analysis examples generally involve a scientific context, even though the concepts and principles apply more broadly. Science is a transparent field where the methods, results, and critiques are all available in journal articles. Conversely, big data for corporations and government institutions is more closed and less freely available. Hence, most of my examples are scientific.

Additionally, I refer to the framework that statisticians developed to understand data quality, experimental designs, and analytical procedures. However, to reiterate, this is not a traditional statistics book because it focuses on the issues you should consider while thinking analytically.

This book is all about arming you with the right mindset to tackle the natural shortcomings we face when dealing with large amounts of information. Whether you're a professional using data in various capacities, a decision-maker eager to be more data-driven, or someone who wants to make better sense of the world without being misled, this book has something for you. I wrote it to help you speak intelligently about data analysis and ask the right questions while steering you clear of the common pitfalls that come with the territory. It's your guide to making sound decisions.

By the end of this book, you will have a critical eye for evaluating data, a thoughtful, analytical approach, and the ability to critique analyses by others in your personal and professional life.

Free Resources for this Book

At several points in this book, I mention files you can download from my website. To download them, go to the URL or scan the QR code:

https://statisticsbyjim.com/thinking-analytically

CHAPTER 1

Cognitive Foundation

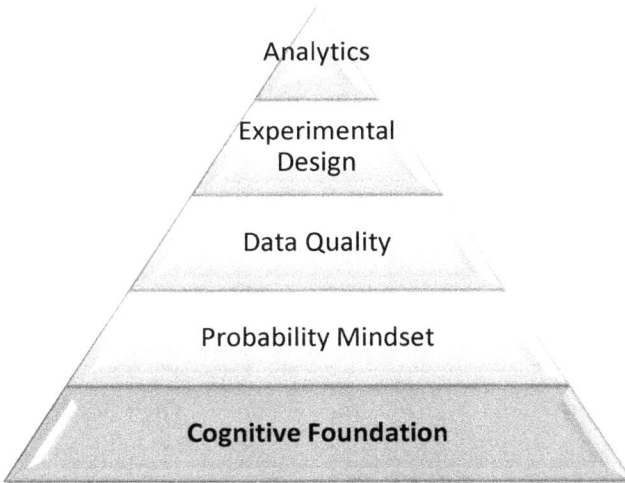

We'll start our journey at the base of the analytical thinking pyramid. This is the human cognitive foundation on which all the other levels build. How good are we at processing information and using it to make unbiased decisions? This chapter answers that question.

In general, everyday life is all about observing information, storing it, and then using it to make decisions about everything from opinions to important life decisions.

We see information all around us, not only in our lives but also in mass amounts in the media. All that information helps us form our political opinions, thoughts about the correct course of action, etc. Our ability to do that logically, critically, and effectively depends on how well our brains think analytically.

So, before getting to data and its analysis, let's explore our brain's capacity for analytical thinking. That might make this chapter feel more at home in a psychology book.

Our brains are the base of the thinking analytically pyramid.

When we devise ways to collect and analyze data, human brains guide every step of that process. And, as you'll see in upcoming chapters, many considerations and complications along the way require human decisions. When you finally get to the analytical results, human brains interpret and communicate them to others.

If you're reading this book, I'll assume that you, like me, consider yourself good at logical analysis. We take pride in it, and some of us even get paid for using that ability in various contexts.

Effective data-based decision-making requires analytical thinking. You must absorb the information, retrieve it when needed, and accurately weigh it to reach a dispassionate, logical conclusion. Our brains are ultimately what all data analysis depends on—the critical, unemotional processing of data to produce valid conclusions.

So, are our brains good at analytical decision-making?

Sure, there's plenty of variation between individuals, but do we effectively process information in an unbiased and accurate manner overall? What is our brainware's capacity? What are its limitations?

Consider this optical illusion. It's known as the Café Wall Illusion because it was initially observed in a café's tile pattern.

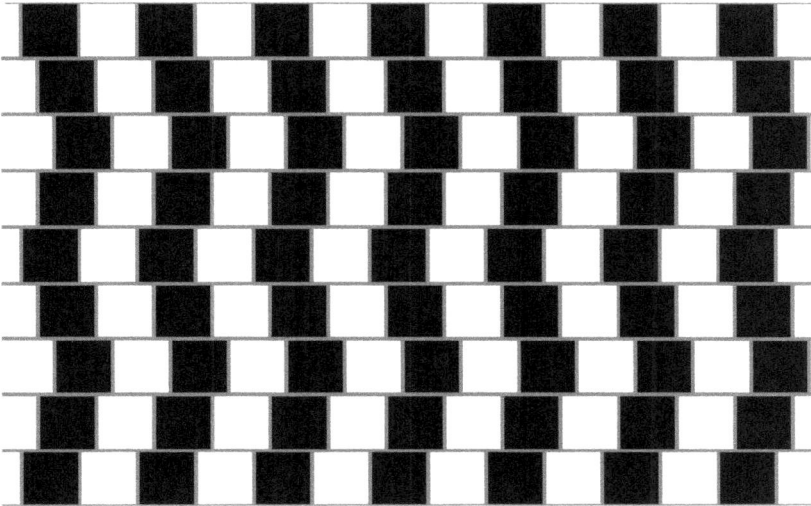

The horizontal lines look like they curve. If you shift your eyes around the image, they might seem to change shape. However, they're all perfectly straight.

Placing a straight edge under one of the unruly horizontal lines temporarily tames it. It looks straight and remains unchanging.

Now, remove the straight edge and, knowing that the horizontal lines are straight, try to overcome the optical illusion by mentally forcing the lines to appear straight.

You can't do it. I want you to ponder the implications of that.

You can't see reality for what it is, even when looking at a simple image.

That should have you questioning your mind and its abilities.

I love optical illusions because you can understand them and how they work, but what you see is inaccurate. They also illustrate how our minds work and force us to acknowledge our limitations.

Our minds often fall prey to optical illusions due to the cognitive shortcuts our brains use to handle the immense amount of visual information we encounter. To efficiently process this information, our brains rely on past experiences and expectations to quickly judge what we see.

This method usually serves us well, allowing us to interpret our surroundings swiftly. However, optical illusions cleverly exploit these shortcuts. They present visual cues that lead our brains to make false assumptions or contradictory conclusions, revealing our visual system's inherent limitations. Optical illusions are a fascinating reminder of how our brain prioritizes efficiency over accuracy in visual processing.

What does this have to do with using data to make decisions?

The same drive to have shortcuts that simplify the deluge of information, causing our brains to fall victim to optical illusions, also causes cognitive biases, which are common errors in thinking. You can think of them as decision-making shortcuts that favor easiness over truth.

Optical illusions and cognitive biases are also analogous in that they can prevent us from seeing the world as it is.

- Optical illusions directly affect how we see reality.
- Cognitive biases affect how we see reality through how we retain, process, and evaluate information.

Cognitive Biases

Research in cognitive psychology, such as the work by Daniel Kahneman and Amos Tversky, has demonstrated that our brains are not naturally wired to process large amounts of data or accurately assess probabilities. These cognitive limitations can lead to systematic biases and errors in judgment.

A cognitive bias is a systematic fault in thinking that can affect our judgments, perceptions, and decisions. These biases arise due to our limited mental capacity, the complexity of the environment, and the influence of our prior experiences and beliefs.

The human brain is a powerful tool but is subject to attentional limits, individual motivations, heuristics, social pressures, and emotions. These factors can all contribute to cognitive biases, many of which are attempts to simplify information processing.

Biases can stem from rules of thumb that help you understand the world and make decisions quickly. But they can also lead to errors and distortions in thinking. For example, limits of attention can lead to incomplete information processing. Individual motivations can circumvent logic, producing biased interpretations. Heuristics, or mental shortcuts, can be helpful in some situations but can also lead to errors if applied inappropriately.

Social pressures can also influence decision-making, with individuals often conforming to the opinions of those around them. Emotions also play a significant role in developing biases. Individuals frequently make decisions using feelings rather than logical analysis. Understanding these factors and their potential impact on thinking can help individuals recognize and mitigate the effects of cognitive biases.

Numerous types of cognitive biases can affect our thinking and decision-making processes. Let's consider some of the critical types of thinking errors.

Confirmation Bias

Confirmation bias is the tendency to seek information confirming preexisting beliefs while ignoring information contradicting them. This bias can be particularly problematic when making important decisions, leading to flawed reasoning and inaccurate conclusions.

Confirmation bias affects how we gather information, as well as how we pay attention, interpret data, and remember information.

Attention: Individuals selectively attend to information confirming their preconceptions while ignoring information contradicting them. For example, someone who believes vaccines are harmful might focus on anecdotal stories of adverse reactions while ignoring large-scale studies demonstrating their safety.

Interpretation: Confirmation bias can cause individuals to interpret information in a way that confirms their preconceptions while disregarding alternative explanations. For example, a person who believes climate change is a hoax might interpret a cold winter as evidence that the planet is not warming while disregarding overwhelming evidence to the contrary.

Recall: Individuals remember information in a way that confirms their preconceptions while forgetting contradictory information. For example, a person who believes a political candidate is corrupt might remember every negative story they read about that candidate while forgetting positive reports contradicting their beliefs.

Confirmation bias is how our brains take shortcuts when we evaluate evidence. Our brains evolved to handle a much slower flow of information than the modern world provides. Historically, people acquired only a little new information during their lives and made most of their decisions for survival. But now, we are constantly bombarded by information from various sources, such as other people, the media, and our experiences.

Evaluating evidence can be challenging, especially when it's complicated, contradictory, or unclear. Constantly assessing and challenging our worldview is a mentally exhausting process. Holding different, opposing ideas in our heads is hard work.

Our brains naturally take shortcuts to keep from feeling overwhelmed. It does this by recalling information that supports our preexisting views and discounting contradictory data. These shortcuts save us time and mental energy. It's much easier to focus on just one worldview and dismiss or ignore any information that challenges it.

This process leads to a confirmation bias feedback loop, where our beliefs become more entrenched over time, making it increasingly difficult to see evidence that doesn't fit our existing views.

So, we prefer to strengthen our worldview by seeking, interpreting, and remembering ideas that support it.

Examples

To better understand these problems, let's look at a few examples.

Political polarization: In today's highly polarized political climate, confirmation bias is a significant issue. Many people seek out news sources that confirm their political beliefs and dismiss those that don't. This practice can lead to a limited perspective on political issues, missed opportunities to consider alternative viewpoints, and a widening political divide.

Medical decisions: Confirmation bias can also impact medical choices. For example, a doctor might be biased towards a particular treatment because they have had success with it previously and ignore evidence suggesting a different treatment is more effective. This disregard can lead to missed opportunities to provide patients with the best care.

Investment decisions: Confirmation bias can also impact investment decisions. An investor might be biased towards a particular stock or market because they have had success with it and ignore evidence suggesting a different investment is more prudent. This approach can lead to missed opportunities to diversify a portfolio and mitigate risk.

Problems

There are a few different types of problems that can arise from confirmation bias:

Limited perspective: When individuals seek information that confirms their beliefs and ignore contradictory evidence, they get a narrow view of the issue. Without a full range of information, it's hard to make informed decisions.

Missed opportunities: Confirmation bias can also lead to missed opportunities, as individuals can overlook important information or ideas that don't align with their preconceived notions.

Polarization: Finally, confirmation bias can lead to polarization, as individuals become increasingly entrenched in their beliefs and less open to considering other perspectives.

Signs

Confirmation bias can be challenging to identify, as it often occurs subconsciously. However, some signs can indicate the presence of confirmation bias, including:

Selective exposure: People actively seek information that confirms their beliefs and avoid information that contradicts them.

Misremembering: People unconsciously alter their memories to fit their beliefs or remember information in a way that confirms their preconceptions.

Cherry-picking: People selectively choose and interpret evidence to support their beliefs while ignoring evidence that contradicts them.

By understanding these symptoms of confirmation bias, individuals can work to become more aware of their own biases and make a more conscious effort to consider alternative viewpoints and evidence.

Tips

Awareness is vital for avoiding confirmation bias. By recognizing the potential for bias in our own thinking, we can be more vigilant in seeking information that challenges our beliefs and preconceptions. Here are a few tips to help avoid confirmation bias:

- **Be open-minded**: Try to approach new information with an open mind, even if it challenges your existing beliefs. Consider alternative viewpoints and be willing to revise your opinions if necessary to reduce confirmation bias.
- **Seek out diverse sources**: It's essential to seek out information from various sources, including those that do not align with your existing beliefs. This practice can help you gain a more balanced perspective and avoid reinforcing your biases.
- **Evaluate the evidence**: To minimize confirmation bias, take the time to evaluate the evidence presented to you and be willing to change your beliefs in light of new information. Avoid cherry-picking evidence that supports your preconceptions, and be skeptical of sources that make extraordinary claims.
- **Be mindful of your emotions**: Our emotions can influence our thinking and lead us to overlook or dismiss evidence that contradicts our beliefs. Be aware of your emotional responses

and try to approach new information with a calm and rational mindset.

Considering these tips and remaining vigilant against confirmation bias can improve your critical thinking skills and help you make better decisions.

Availability Heuristic

The availability heuristic is a cognitive bias that causes people to rely too heavily on easily accessible memories when estimating probabilities and making decisions. This mental shortcut can distort our perception of how frequently certain events occur.

A classic example of the availability heuristic is believing that airplanes are unsafe because of highly publicized plane crashes. This bias contradicts statistical evidence showing that flying is a much safer mode of transportation than driving.

Have you ever made a decision based on information that was readily available in your memory? If so, you might have fallen prey to the availability heuristic.

The availability heuristic is a mental shortcut that our brains use to evaluate probabilities rapidly. When we face a decision, we rely on information that comes to mind quickly. This includes information we can recall more easily, events that affected us emotionally, and recent events.

Unfortunately, this cognitive bias distorts our ability to judge the probability of events accurately. Our memories frequently are not realistic models for forecasting future outcomes.

For example, imagine estimating the likelihood of a car accident on your daily commute. Your brain might rely on readily accessible information, such as past experiences or media coverage, rather than statistical data, leading to inaccurate estimates.

Some examples of the availability heuristic include:

- Assuming that all sharks are dangerous and that shark attacks are common because of media coverage of shark attacks. In reality, the vast majority of shark species are not harmful to humans. Indeed, there were only 57 confirmed shark attacks worldwide in 2022.
- Overestimating the risk of being a victim of a terrorist attack. Statistically, the likelihood of being killed in a terrorist attack is extremely low.
- Assuming that all large dogs are aggressive and dangerous because of news stories or personal experiences with them.

Psychologists Amos Tversky and Daniel Kahneman coined the term "availability heuristic" in 1973. They suggested that it operates subconsciously and uses the principle, "If you can think of it, it must be important." This notion leads people to believe that things that come to mind more quickly are more common.

Information you know *and* comes to mind easily

All information

The availability heuristic is based on ease of retrieval. The more easily accessible information is, the more likely people are to rely on it to evaluate probabilities and make decisions. Media coverage, recency, and the unavailability of information can foster the availability heuristic.

Media Coverage

Media coverage is one of the main factors affecting the availability heuristic. When news outlets or social media expose us to particular events, they become more salient in our minds. For example, if we hear about a plane crash, we might become more fearful of flying, even though flying is one of the safest modes of transportation.

Sensationalized news stories such as reports of shark attacks, plane crashes, and child abductions instigate fear. Incidents of this magnitude are splashed all over the media and can create hysteria. This media coverage leads us to believe these catastrophes are more common than they are.

Recency

Recency is another factor that can influence the availability heuristic. When something has happened recently, it is more likely to be at the forefront of our minds. For example, if you have just seen a car accident on your way to work, you might be more likely to take a different route, even though the accident is uncommon.

Unavailable Information

The unavailability of information can also contribute to the availability heuristic. When we cannot access certain information, we must rely on what we know, even if it is incomplete or inaccurate. For example, imagine trying to estimate the percentage of people in your country who are immigrants. You rely on your personal experiences and interactions, even though they probably don't represent the overall population.

Awareness of this bias can help us make more informed decisions and avoid judgments based on incomplete or inaccurate information. Keep in mind that it's not just what you know but the subset that comes to mind quickly.

The way we see the world is different from how it is. Our expectations about how often things happen are skewed by how frequently we hear about them and how strongly we feel about those messages.

Representativeness Heuristic

The representativeness heuristic is a cognitive bias that occurs while assessing the likelihood of an event by comparing its similarity to an existing mental image. Essentially, this bias involves comparing whatever we're evaluating to a situation, prototype, or stereotype that we already have in mind. Our brains frequently weigh this comparison much more heavily than other relevant factors. This shortcut can be helpful in some cases, but it can also lead to errors in judgment and distorted thinking.

Have you ever made a snap judgment about someone based on their appearance or personality? This type of assessment exemplifies the representativeness heuristic.

When many people imagine a physics professor, they might picture an older person with messy hair and rumpled clothes. If someone in shorts and a Hawaiian shirt introduces themselves as your new physics professor on the first day of class, you might feel taken aback. This reaction occurs because they do not fit your mental prototype of a physics professor.

While the representativeness heuristic offers rapid decision-making, it can cause you to disregard crucial information and reach erroneous conclusions. In the extreme, it leads to stereotyping and prejudice.

The following examples illustrate the distortions that can occur when we rely on stereotypes and generalizations to make judgments and decisions.

All the following are examples of the representativeness heuristic. You assume that a person:

- Wearing a business suit is wealthy and successful, even though they could have borrowed it from a friend for a job interview.
- With a tattoo and piercings is rebellious and anti-establishment, even though they might enjoy expressing themselves through body art.
- Driving a luxury car is wealthy, even though they might have bought it used or on lease.
- Wearing glasses is intelligent.

Imagine meeting two people for the first time, neither of whom you've met. One of them is an artist, and the other is a scientist. You don't know anything else about them. When you meet them, you see that one is wearing glasses and has a calculator in their pocket. The other is wearing a beret. Based on this information alone, you might assume that the person with the glasses and calculator is the scientist and the person in the beret is the artist. However, this is a clear example of the representativeness heuristic in action, and it's essential to recognize and avoid it.

You know you shouldn't judge a book by its cover. However, the representativeness heuristic causes many to do just that! I've published three books before this one, and all the expert advice I received said that you must have a good cover, or people won't buy it!

Psychologists Amos Tversky and Daniel Kahneman first discovered the representativeness heuristic in the 1970s. They found that people often rely on stereotypes and generalizations when making judgments and decisions because it is quicker and easier than considering all the relevant information. In other words, our brains take shortcuts when processing data.

When determining the probability that an object X belongs to a category Y, our brain often relies on the representativeness heuristic. Essentially, we rely on the perceived similarity between X and Y to make this judgment, frequently giving it higher weight than more relevant factors.

For instance, if X has certain qualities typical of Y, we might assume that X must belong to Y, even if there are other important factors to consider. The representativeness heuristic relies on similarity instead of more intricate probabilistic and logical explanations.

We rely on the representativeness heuristic due to our limited cognitive resources. Every day, we face thousands of distinct decisions, and our brains make them while conserving as much energy as possible. This heuristic focuses on categories and our tendency to overemphasize similarities.

Categories and Mental Prototypes

Categorization is crucial to comprehending the world around us. Although it might appear obvious, categories play a more fundamental role in our ability to function than many realize. Think of all the diverse things you encounter in a single day. Whenever we interact with objects, animals, or people, we rely on our knowledge of their category to determine what to do.

For instance, when we visit a park filled with various bird species, we can categorize them all as "birds," which allows us to expect that they can fly and lay eggs and that we should avoid disturbing their nests.

We depend on categories to make sense of the world around us. Without categories, every time we encounter something new, we'd need to learn about it from scratch, which would be impractical due to our limited cognitive capacity.

For instance, we can recognize a car upon seeing one, even when we don't know the exact make and model. Intuitively, we know what to expect.

By grouping similar items, we draw on our knowledge of the category and can immediately take appropriate action. That's the positive side of the representativeness heuristic.

According to prototype theory in psychology, we construct these categories around prototypes representing typical category members. Consequently, these prototypes seem more representative of that category than others. Prototypes influence how we perceive category members by serving as a basis for comparison. Thus, they are a critical component of the representativeness heuristic.

For example, penguins are birds, but they feel "off" because they don't fit our mental prototype of a bird as well as a robin. That's the downside of the representativeness heuristic.

Overemphasize Similarity

We tend to overemphasize similarity and ignore more pertinent information. Dependence on similarity leads people to overlook "base rate" information about an event's frequency. For instance, when asked to categorize a person as a programmer or an artist, most people would immediately categorize that person as a programmer after being informed that the person enjoys coding—even if they knew that the person comes from a population containing only 10% programmers.

In the next chapter, we'll explore the challenges of accurately incorporating base rate information into our decision-making in much more detail.

How to Avoid the Representativeness Heuristic

The representativeness heuristic can lead to biased thinking and errors in judgment. Understanding our tendency to rely on stereotypes and generalizations when making judgments and seeking and considering all relevant information is essential. By doing so, we can avoid making inaccurate judgments and decisions based on stereotypes and generalizations.

Now that we understand the representativeness heuristic and the problems it can cause, how can we avoid falling into this cognitive trap? Here are a few strategies:

Look for Base Rate Information

Take the time to gather information about how often certain events occur in general, not just in the specific example you are considering. This process can help you avoid making snap judgments based largely on similarity.

Be Mindful of Your Mental Prototypes

Be aware of the prototypes and stereotypes you use when making judgments. Try to approach each situation with an open mind and avoid jumping to conclusions based on surface-level similarities.

Use Statistical Thinking

Try to think probabilistically and logically when making judgments rather than similarity. The representativeness heuristic short-circuits these aspects of mental evaluation.

Seek Out Diverse Perspectives

Exposure to various perspectives and experiences can help challenge your assumptions and reduce reliance on stereotypes and prototypes.

Seek out diverse viewpoints and actively try to understand different perspectives.

Slow Down

The representativeness heuristic often occurs when we make quick, snap judgments without taking the time to consider all available information carefully.

By being aware of this heuristic and using these strategies to avoid it, you can make more informed and accurate judgments.

Anchoring Bias

Anchoring bias is a cognitive bias that causes people to rely too heavily on the first piece of information they receive when making a decision. That information is their "anchor." Even when presented with additional information, people tend to give too much weight to the original anchor, leading to distortions in judgment and decision-making.

For example, imagine you're shopping for a new laptop. The first one you see costs $2,000, and the price shocks you. This initial price is your anchor. Later, you come across another laptop that costs $1,500, which seems like a good deal compared to the $2,000 one. However, if you had seen the $1,500 laptop first, you might not have thought it was such a great deal because you wouldn't have had the $2,000 price for comparison.

Anchoring bias can have significant consequences when objective decision-making is critical, such as in negotiations or financial planning.

In the medical field, anchoring bias can have serious consequences. Doctors might rely heavily on initial symptoms rather than subsequent ones when making a diagnosis.

For instance, the COVID-19 pandemic has raised concerns about the impact of anchoring bias on patient diagnoses. A study conducted in 2020 found that doctors can be prone to anchoring bias when diagnosing COVID-19, potentially leading to failure to diagnose other conditions (Yousaf et al., 2020).

Psychologists Amos Tversky and Daniel Kahneman first discovered the phenomenon in the 1970s. They proposed that anchoring bias occurs because the anchor serves as a reference point for subsequent judgments. People use the anchor as a starting point and then adjust their decisions from there rather than starting from scratch with each new piece of information. That fits the common theme of efficiency over accuracy.

In Tversky and Kahneman's anchoring bias study, participants estimated the percentage of African countries that were members of the United Nations. However, before answering the question, participants were asked to spin a wheel rigged to land on either 10 or 65. Participants who landed on 10 gave estimates much lower than those who landed on 65, even though the number on the wheel was irrelevant to the actual percentage of countries in the UN.

Psychology theories suggest that anchoring bias occurs because it's a shortcut our brains use to make quick decisions. Rather than carefully weighing all the available information, our brains latch onto the first piece of information we encounter to streamline the decision-making process.

Research suggests that mood can impact anchoring bias, with individuals with a positive attitude being more likely to adjust correctly from the anchor than those in a negative mood (Englich & Soder, 2009). Similarly, personality traits such as openness to experience tend to mitigate the effects of anchoring bias. Individuals high in openness tend to be more receptive to new information and less likely to rely solely on the initial anchor (Caputo, 2014).

Experience can also play a role in anchoring bias. In one study, participants with expertise in a particular subject area were less susceptible to it when making estimates related to that domain (Welsh et al., 2014). Consequently, experience and knowledge can help individuals overcome the effects of anchoring bias.

While anchoring bias can lead to distorted judgments and decision-making in all individuals to some degree, factors such as mood, personality, and experience can influence the extent to which individuals are susceptible to this cognitive bias. By understanding these factors and working to mitigate their impact, individuals can avoid the pitfalls and make more objective decisions.

Framing Effect

The framing effect is a cognitive bias that distorts our decisions and judgments based on how information is presented or "framed." This effect isn't about lying or twisting the truth. It's about the same cold, hard facts making us think and act differently just by changing their packaging.

Have you ever noticed how the same facts presented differently can lead to entirely different decisions?

Framing is all about how we present options. Are they shown as a win (positive frame) or a loss (negative frame)? We naturally choose options that look like gains, even if a 'loss' option leads to the same outcome.

Think about it. Would you opt for surgery if the surgeon says it has a 90% survival rate? Probably yes. But would you feel the same if they said it has a 10% mortality rate? Suddenly, it sounds riskier, right? Statistically, these are the same outcomes, but they feel different. That's the framing effect in action!

The following are examples of the framing effect that you've probably seen:

- **Marketing Tactics**: Notice how advertisements say "Save 25%" instead of "Spend 75%." This positive framing makes the deal seem more attractive.
- **Food Choices**: A label saying "99% fat-free" is more appealing than one stating "contains 1% fat," even though they mean the same thing.
- **Political Campaigns**: Politicians often use framing to their advantage. For instance, a policy might be presented as "protecting national security" instead of "increasing surveillance," even though both statements refer to the same action. The first frame appeals to our desire for safety, while the second highlights potential privacy concerns.

Kahneman and Tversky identified the framing effect while laying the groundwork for what we now know as prospect theory.

Prospect theory is a psychological model that describes how people make decisions when faced with risk. It fundamentally challenges the traditional economic idea that humans are rational actors who always make decisions in their best interest. Instead, Kahneman and Tversky found that people often make irrational decisions influenced by how choices are framed.

A key component of prospect theory is the idea of loss aversion – we hate losing more than we enjoy winning. Studies suggest that the pain of losing is psychologically about twice as powerful as the pleasure of gaining. So, when choices are framed as potential losses, we tend to take more risks to avoid those losses. Conversely, we tend to be more conservative when the same options are stated as potential gains.

Take the surgery example we discussed earlier. The statement "90% chance of survival" is framed as a gain, which makes us prefer the safer, more conservative option. On the other hand, a "10% chance of death" is a potential loss, making us more likely to avoid the risk associated with it.

Another reason the framing effect occurs is our reliance on mental shortcuts. Our brains take these shortcuts to save time and cognitive effort. But these shortcuts can also lead us astray, especially when the framing of information nudges us toward a particular choice.

The framing effect, deeply intertwined with prospect theory, is a potent force that can sway our decision-making process. It's a reminder that our decisions are not always as objective or rational as we might like to think. Awareness of this cognitive bias can help us make more informed and balanced decisions in our daily lives.

Here are a few strategies to help dodge the framing effect in our everyday lives:

- **Awareness is Key**: Knowing about the framing effect is a significant first step. When you're aware, you can notice how the framing of information influences your decisions.
- **Reframe the Information**: When faced with a decision, try reframing the information differently. If a situation is framed positively, think about how it would look if stated negatively, and vice versa. This process can help you see the choice more objectively.
- **Seek More Information**: Don't rely solely on how information is presented. Dig deeper and gather more data. The more information you have, the less likely you'll be swayed by how it's framed.
- **Take Your Time**: Impulsive decisions are often where cognitive biases sneak in. If you can, take your time to make decisions. Slowing down allows you to consider the information more carefully and reduces the influence of the framing effect.

By applying these strategies, we can reduce the power of the framing effect and make decisions that align with our true preferences and values.

Hindsight Bias

Hindsight bias is a cognitive bias that causes people to perceive past events as more predictable than they actually were. It is that sneaky feeling that you "knew it all along," even when that's not true. This tendency is rooted in our desire to believe that we are intelligent and capable decision-makers, which can cause various distortions in our thinking.

Have you ever felt that events were inevitable after they happened?

Perhaps you've watched a movie with a twist ending and afterward felt like it had to end that way. Or maybe after watching a sporting event or tracking stocks, you felt like you always knew the outcome.

Hindsight bias makes events appear more obvious and predictable in retrospect. It can lead us to believe that we should have seen something coming, even if it was impossible to predict.

Hindsight bias is our brains processing information in a way that makes us feel good about ourselves. When we know the outcome of an event, our brains automatically assume that we would have predicted that outcome all along, even if we had no information to support that prediction beforehand. After all, recognizing that some outcomes are unpredictable can make us feel uneasy.

Unfortunately, hindsight bias can distort our perceptions.

One such distortion is overconfidence. By believing we knew an outcome all along, we become overly confident in our ability to predict future events. This problem can lead to poor decision-making and risky behavior.

Hindsight bias also causes us to discount the role of randomness in events by making us think the outcome was inevitable. This belief can lead us to make incorrect assumptions about the likelihood of future occurrences. Understanding the role of randomness and variation in outcomes becomes major points later in this book.

Psychologists are not immune to the effects of hindsight bias. Clinical psychologist Paul E. Meehl noted during a conference that clinicians often overestimate their ability to predict the outcome of a particular case and claim to have known it all along.

A study by Dorothee Dietrich and Matthew Olson asked college students to predict the outcome of the U.S. Senate vote on the confirmation of Supreme Court nominee Clarence Thomas. Before the vote, 58% of participants predicted he would be confirmed. After the confirmation, 78% claimed they had predicted it. This discrepancy demonstrates the effect of hindsight bias on our perception of past events.

Psychologist Baruch Fischhoff identified hindsight bias in 1975. In this study, Fischoff compared participants who did and did not know the results of obscure historical and clinical events. Subjects who the researchers told the outcomes were more likely to report having predicted that outcome than those who weren't told. Crucially, participants were unaware of how knowing the outcomes affected their perceptions and tended to overestimate what they would have known without this information.

Fischoff concluded that when we look back on events, we often see them as logical and predictable, unfolding regularly and linearly with an inner sense of necessity, making us feel they couldn't have happened any other way. He referred to this as "creeping determinism."

Since then, psychologists have proposed other theories to explain why it occurs.

One theory is that hindsight bias results from our brain's natural tendency to organize and simplify information. When we know the outcome of an event, our brains reorganize the information to make it seem more straightforward and predictable than before the event occurred.

Another theory is that hindsight bias results from our need to maintain a positive self-image. When we believe that we knew the outcome of an event all along, we are more likely to feel good about ourselves and our abilities, even if those beliefs are not based on reality.

Hindsight bias is a sneaky cognitive bias that can distort your thinking and decision-making. By being aware of its effects and taking steps to combat it, you can make more informed decisions and avoid overconfidence in your predictive abilities.

One way to combat hindsight bias is to keep a journal of your predictions and their reasoning. By comparing your past predictions to the actual outcomes, you can better understand our thought processes and biases.

Additionally, it's crucial to recognize the role of chance and randomness in events. It is tempting to attribute outcomes to your predictive abilities. However, you must acknowledge that luck and other external factors play vital roles in the results.

Remember, just because something seems obvious in hindsight doesn't mean it was predictable before it happened. We should strive to make decisions based on the information available to us at the time and avoid being overly confident in our ability to predict future events.

Dunning-Kruger Effect

The Dunning-Kruger effect is a cognitive bias that causes people with low abilities or knowledge to overestimate themselves compared to others. Conversely, people with high skills tend to underestimate themselves. In short, it is a psychological phenomenon that distorts our self-evaluation.

You've probably seen old high school friends on social media or had dinner guests who are "experts" on various subjects, which seem to change weekly. They'll rattle on confidently about complex topics that they know little about while being blatantly unaware of their factual shortcomings. And if you try to correct them factually, watch out! That's the Dunning-Kruger effect in action.

The Dunning-Kruger effect influences people who lack the necessary skills and knowledge to evaluate their performance accurately. This deficiency creates a dangerous combination of poor self-awareness and limited abilities, leading them to think they are better than they actually are. Overestimating their abilities can lead to mistakes, poor decisions, and resistance to constructive feedback or criticism.

Interestingly, it can also affect highly qualified and intelligent people. Research has shown that people with knowledge in a particular area might underestimate their abilities relative to others.

High performers tend to compare themselves to even more qualified people, making them feel like they don't measure up. Additionally, they are often keenly aware of how much they don't know, causing them to feel less confident. Thus, even highly qualified individuals can benefit by recognizing the Dunning-Kruger effect.

Examples

The Dunning-Kruger effect can affect people from all walks of life, regardless of age, background, or education.

Driving

The Dunning-Kruger effect is in action on the road. Studies have found that less experienced drivers overestimate their driving abilities. For example, a new driver might believe that they are excellent and take risks that a more experienced driver would avoid. This overconfidence can lead to accidents and other dangerous situations on the road.

Sports

The Dunning-Kruger effect also exists in sports. Less skilled athletes overestimate their abilities and make mistakes that cost their team the game. This problem is particularly evident in team sports like basketball, where players believe they are better than they actually are and try to make plays beyond their skill level. This overconfidence can result in turnovers and missed shots, ultimately hurting their team's chances of winning.

Work and Career

In the workplace, the Dunning-Kruger effect can manifest in various ways. For example, a new employee might believe they understand the job requirements and not ask enough questions or seek feedback from their supervisor. This overconfidence can lead to misunderstandings and mistakes that they could have avoided. Studies have shown that the poorer performers typically think they're above average. Hence, they do not seek out training and assistance.

Highly skilled and knowledgeable employees can also experience the Dunning-Kruger effect. These employees might have been in the same position for so long that they underestimate their abilities and assume others possess the same expertise. This problem can lead to a lack of self-confidence, preventing them from taking on new challenges or pursuing opportunities.

Is It Real?

Unfortunately, yes, it exists. In addition to your own annoying experiences with those suffering from this condition, scientific research supports the existence of the Dunning-Kruger effect.

This effect is named after David Dunning and Justin Kruger, who first described the phenomenon in a 1999 paper. Their research showed that people with low knowledge in a particular area often have insufficient knowledge to recognize their limitations. As a result, they tend to overestimate their abilities and make mistakes without realizing it.

How does this happen?

In this context, the concept of the "twin burden" refers to the lack of knowledge or skills that both causes poor outcomes *and* prevents people from accurately evaluating their performance. In other words, they cannot perform well, and the same lack of knowledge distorts their self-assessment.

They don't know what they don't know.

Consequently, they think they're doing better than they actually are. This bias can lead to overconfidence, a lack of motivation to learn or improve, and poor decision-making.

To measure the Dunning-Kruger effect, researchers ask people to take a knowledge or abilities test and then ask them how well they think they did. Researchers compare the test scores to the self-evaluations. Frequently, the lowest performers see themselves as above average.

Research shows that low scorers on grammar, humor, and logic tests often overestimate their performance significantly. In one study, participants in the lowest percentiles on these tests believed their performance was much better than it was. For example, those below the 12th percentile estimated they performed at the 62nd percentile, demonstrating a significant gap between perception and reality.

The adverse effects of the Dunning-Kruger effect exist at both an individual and societal level.

On a personal level, people who overestimate their abilities might take on tasks they're incapable of completing, leading to mistakes and failures. They tend to resist feedback and criticism, believing they already know everything they need to know. These problems can lead to stagnation in personal and professional growth.

On a societal level, the effect contributes to the spread of misinformation and conspiracy theories. People who lack knowledge about a topic might feel overly confident about forming accurate opinions about it. This problem can lead to people spreading false information or rejecting scientific consensus, which can negatively affect public health and safety.

Avoiding It

It seems like the Dunning-Kruger effect has worsened with the growth of the Internet, quick Google searches, and YouTube videos. It's so easy to obtain some knowledge and feel like you've become knowledgeable. Using this approach, it can seem like we know all the relevant details of a topic, but that pales in comparison to someone who has spent a career or lifetime studying the area.

So, how can you avoid falling victim to the Dunning-Kruger effect? The first step is recognizing that self-evaluation is tricky and prone to errors. Consequently, simply being aware of it can help you avoid over or underestimating your abilities. You can look for the signs.

If you think you're bad at something, it likely means you have some insight into your limitations. That's a good thing.

On the other hand, if you think you're an expert, remain humble and recognize there is always room for improvement.

Am I new to this subject area? If yes, I'm probably not an expert or even above average!

Am I rejecting criticism out of hand?

Low performers tend to struggle with criticism and resist self-improvement. Still, by embracing feedback and using it mindfully, you can move forward and avoid the negative consequences of the Dunning-Kruger effect.

Don't be afraid to seek feedback and criticism; be open to learning and growing. Remember, there's always more to learn. Recognizing your limitations is the first step to becoming more knowledgeable and competent.

Closing Thoughts

The common causes of various cognitive biases are rooted in the distortions of how we process, retain, and utilize information. These biases arise because our cognitive system takes mental shortcuts for efficiency, leading to errors. These causes include the following traits:

- Paying attention to and remembering information that aligns with our existing beliefs while ignoring or forgetting contradictory data.
- Relying on immediate examples that come to mind when evaluating a topic often leads to overemphasizing recent or emotionally charged information.

- A propensity to make judgments based on how much something resembles our mental prototypes, sometimes leading to misjudgment of actual probabilities.
- Relying heavily on the first piece of information we receive can skew our subsequent thinking and decision-making.
- Influenced by the way information is presented rather than the information itself.
- Emotions affect how we weight information and how likely we are even to recall it.

In short, when making objective decisions, we often remember only some information influenced by our prior beliefs, recent events, and emotionally charged messages. We then evaluate this biased information based on the order in which we received it, its similarity to our mental models, how it was presented, and how emotions weighted it.

This combination of biased information and poor evaluation methods are not ideal. However, there are ways to combat these cognitive biases.

Critical strategies to mitigate them involve cultivating an awareness of these biases so we can question our thought processes. Practical approaches include:

- Seeking diverse perspectives.
- Relying on objective data over anecdotes.
- Considering multiple viewpoints before decision-making.
- Maintaining a critical approach toward sources and contexts of information.
- Keeping accurate records of past decisions.

Humans are not great at objectively recalling and evaluating information to make decisions in everyday life with its messy information overload. We shouldn't be surprised because our brains run on only 20 watts of power. While it's impressive how much our minds can do with such a limited amount of energy, they must find ways to conserve it. Hence, the shortcuts and simplifications I've discussed throughout this chapter.

The bizarreness effect, a concept brought to light by McDaniel and Einstein, suggests something interesting: we're more likely to remember information if it's, well, bizarre. This effect doesn't mean every piece of data needs to be outlandishly strange, but it does highlight how our brains perk up at the unusual or unexpected. The oddities and anomalies in the information we encounter often stick with us the most easily. So, what you recall from a sea of information are often the remarkable bits. In short, unusual information affects us more than the usual stuff, giving it an outsized impact on our opinions and decisions.

The pioneering work of Daniel Kahneman and Amos Tversky fundamentally challenged the notion of humans as rational beings. By investigating various cognitive biases, they demonstrated that systematic errors and irrational tendencies often influence our decision-making processes. Their research revealed that, contrary to the traditional economic theory that portrayed humans as logical and optimal decision-makers, we frequently rely on heuristics and are subject to biases that distort our judgments. This shift in understanding has profound implications for humans as analytical thinkers.

Remember the optical illusion at the beginning of the chapter? Sometimes, you can't see reality for what it is due to limitations built into your brain.

Similarly, cognitive biases make it hard to see reality clearly when making data-based decisions. These biases act like mental illusions, leading us to assign incorrect weights to outcomes that don't align with their actual probabilities. Kahneman and Tversky's work highlights these cognitive distortions, showing how our perceptions often stray from the truth.

In this book, I encourage you to become a skeptic, especially regarding things that hit close to home. Question everything—others' claims, yes, but also your own beliefs. Ask yourself, "Why do I hold these beliefs? Are my data solid? Where did my information come from?"

It's crucial to sidestep common pitfalls by ensuring you use broadly sourced data, properly weight all data points without letting emotions like fear or desired outcomes cloud your judgment, and remember to consider both the data you have and the data you might be missing.

Always be extra cautious with results that seem to neatly confirm what you already believe—this can be a red flag that you're looking at things through a biased lens.

I've highlighted cognitive biases for several reasons.

1. Being aware of these cognitive biases and using the tips for overcoming them will help you be a better analytical thinker—the primary goal of this book.
2. Later chapters build on this information to show how analytical methods try to counteract these cognitive errors and how they affect analysts while performing analyses.

This is our cognitive foundation, on which everything else is built. Are we better at handling more controlled situations with defined probabilities? Let's see!

❊

CHAPTER 2

Probability Mindset

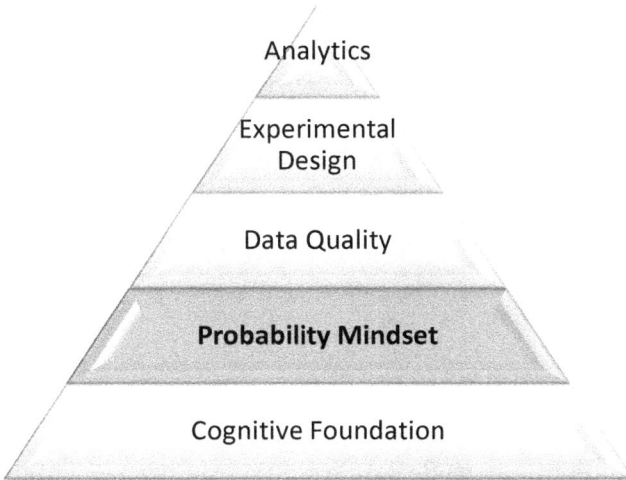

The previous chapter explored the human brain's limitations when re-taining and using information to understand the world, form opinions, and make decisions. We didn't pass with flying colors. And that's the base level of our analytical thinking pyramid!

But perhaps the world is just too complex. When given specific infor-mation, we might be better at solving bite-sized problems.

Let's see if we're better at the next level of the Thinking Analytically Pyramid level—Probability Mindset. This level is where we take specific information to solve a given problem. Calculating probabilities are crucial for finding making decisions that are more likely to produce good results.

When real-world problems involve simple math, we can do okay. However, sometimes even those can be tricky. Think back to algebra class and word problems. They were doable, but they often required some thought.

In data analysis, you frequently work with probabilities and evaluate the likelihood of different outcomes to help you make decisions. Consider the following examples:

- Estimating the success rate of a new marketing campaign by analyzing historical sales data and customer feedback.
- Predicting the probability of catastrophic rainfall in a specific region using climate models and historical weather patterns.
- Calculating the odds of a patient responding to a new medical treatment based on clinical trial results and patient demographics.
- Determining the risk of investment in a startup by examining market trends, competitor success rates, and economic forecasts.
- Assessing the chance of a cyber-attack on a company's network by analyzing past security breaches and current threat levels in the industry.

Are we good at understanding probabilities and using them to make decisions? That's what we'll cover in this chapter.

I'll start with a brief overview of probabilities. This chapter isn't about all the various calculation methods because that would fill up an entire book! But I'll cover some of the basics. Then, I'll move into several probability puzzles and additional cognitive biases relating to misunderstanding likelihoods.

Probability Overview

The definition of probability is the likelihood of an event happening. Probability theory analyzes the chances of events occurring. You can think of probabilities as being the following:

- The long-term proportion of times an event occurs during a random process.
- The propensity for a particular outcome to occur.

For example, we're all familiar with flipping a coin and the chances of getting a "heads" are 0.5. We can apply that to a single coin flip or consider it the long-term proportion of flipping coins many times. We'd expect 50% of all coin flips to produce heads, and there is a 50% chance the next coin flip will be heads.

Probability values range from 0 to 1. Zero indicates that the event cannot happen, while one represents an event guaranteed to happen. Values between 0 and 1 denote uncertainty over whether the event will occur. As the likelihood increases, the event becomes more likely. The middle value of 0.5 signifies that the event is equally possible to happen or not. In a coin flip, the probability of heads occurring equals the likelihood of it not occurring (tails).

What are the chances of *that* occurring?! Have you ever asked yourself that after an unusual occurrence? You can use probabilities in many facets of your personal life. What are the chances of winning the lottery or being in a car accident? Are you more likely to be hit by lightning or winning the lottery? Does wearing a seatbelt change the probability of being injured? How likely is it that you'll become pregnant?

Risks are the chances of bad events, and modeling them is crucial for planning. Actuarial sciences and financial analysts need to understand the likelihood associated with risks to plan for them. Governments use probabilities to know how likely adverse events are to occur and to plan accordingly. How often do catastrophic floods or hurricanes happen in a particular area? What is the likelihood of flood water exceeding a specific level?

Manufacturers must understand the probability of their products' failure over time to avoid unhappy customers and determine their warranties' lengths. Have you ever had a warranty expire just before a product failed? That's no coincidence! At the statistical software company where I used to work, we had a specific "Warranty Analysis" portion of the software that used reliability and failure time data to predict and minimize warranty costs.

You can use probability theory to help you win games of chance by understanding the likelihood of certain outcomes. Unfortunately for gamblers, casinos use probabilities to ensure they'll make profits. The house always wins in the long run!

Statistical hypothesis testing uses probabilities to help you evaluate hypotheses relevant to your study. P-values are a well-known type of likelihood, allowing you to determine whether your results are statistically significant. Probabilities are an integral part of experiments, statistical analyses, and making data-driven decisions.

For example, is the likelihood of contracting the flu lower if you are vaccinated?

Calculations

I'll show you how to calculate simple probabilities to help you understand the fundamentals. We'll look at independent random events where the occurrence of an event, or lack thereof, does not affect future probabilities. For example, the outcome of one coin toss does not affect the outcome of future coin flips.

At its most basic, the probability of an event occurring equals the following:

$$Probability(Event) = \frac{Number\ of\ ways\ event\ occurs}{Total\ number\ of\ outcomes}$$

The numerator equals the number of ways an event can occur. We define what counts as an event based on our interests. For example, we can consider heads in a coin toss or drawing a king from a deck of cards as events. If we define an event as rolling a 1 or 6 on a die, there are two ways an event occurs.

The denominator represents the number of possible outcomes. The subject matter defines this value. For example, coin tosses can have only two results: heads or tails. There are 52 cards in a standard deck of cards. Each outcome is mutually exclusive from the others.

The law of large numbers states that as the number of trials (i.e., coin flips, rolls of the die, drawing cards, etc.) increases, the observed proportion will converge on the expected probability.

Let's start simple with a coin toss and define heads as the single outcome that counts as an event. There is only one way an event can occur and two possible outcomes.

P(H) = 1/2 = 0.5.

I wrote that using standard notation and it indicates that the likelihood of heads equals 0.5.

Now, let's calculate the probabilities for rolling a die. We'll find the likelihood of rolling a 6, a 1 or a 6, and rolling an even number. Notice how each example changes the number of outcomes that count as an event in the numerator. For a standard die, there are always six potential outcomes. Consequently, the denominator is always 6.

- P(6) = 1/6 = 0.167

- P(1 or 6) = 2/6 = 0.33
- P(Even) = 3/6 = 0.50

Finally, we'll calculate likelihoods for a randomized, full deck of cards. What are the chances of drawing any card with a heart (H), any king (K), and a king of hearts (KH)? The denominator indicates the 52 cards present in a full deck.

- P(H) = 13/52 = 0.25
- P(K) = 4/52 = 0.077
- P(KH) = 1/52 = 0.019

However, these chances only apply to the first draw from a full deck. Any card we remove affects the likelihood of the next card. Drawing successive cards from a deck are dependent events, unlike coin tosses and dice rolls. Each event affects the probability of the next one.

Here are some of the other methods and considerations for finding probabilities:

Independent vs. Dependent Probabilities: This distinction is crucial in probability analysis; independent probabilities are where one event does not affect the other, while dependent probabilities involve events that have some degree of dependence on each other. This determination affects how you calculate different types of probabilities.

Conditional Probabilities: This involves calculating the probability of an event occurring, given that another event has already happened, which is essential in scenarios where outcomes are interconnected.

Joint Probabilities: This method deals with finding the likelihood of two or more events happening simultaneously, helping to understand the relationships between multiple variables.

Using Permutations and Combinations to Calculate Probabilities: This approach uses mathematical formulas to determine the probabilities of various outcomes in situations where the order of events matters (permutations) or does not matter (combinations).

That's a far from complete overview of probabilities, but I'm hoping to impress upon you that they're crucial for making decisions and that multiple considerations are involved when calculating them. They're a bit tricky, and honestly, calculating them for complex scenarios can make my head spin despite being familiar with the various methods.

It probably won't surprise you that people generally are not good at understanding probabilities. To illustrate that, I'll start with several probability problems and then cover several general types of errors.

Monty Hall Problem

Who would've thought an old TV game show could inspire a statistical problem that has tripped up mathematicians and statisticians with Ph. Ds? The Monty Hall problem has confused people for decades. In the game show Let's Make a Deal, Monty Hall asks you to guess which closed door a prize is behind. The answer is so puzzling that people often refuse to accept it!

The Monty Hall problem's baffling solution reminds me of optical illusions, where you find it hard to disbelieve your eyes. I think of this puzzle as a statistical illusion. Like an optical illusion, even after knowing the truth, it's hard to see and accept the correct answer, and the illusion can seem more accurate than the actual answer. This statistical illusion occurs because your brain's probability evaluation process uses a false assumption in solving it.

To see through this statistical illusion, we must carefully break down the Monty Hall problem and identify where we're making incorrect assumptions. This process emphasizes how crucial it is to check that you're satisfying the assumptions of a data analysis before trusting the results.

Here's how the game works. Monty randomly hides a prize behind one of three doors, while the other two have no prize.

Then he follows this order of events:

1. Monty Hall asks you to choose one of three doors. You state out loud which door you pick, but you don't open it immediately.
2. Monty opens one of the other two doors that does not have a prize behind it.
3. At this moment, there are two closed doors, one of which you picked. The prize is behind one of the closed doors, but you don't know which one.
4. Monty asks you, "Do you want to switch doors?"

Most people assume that both doors are equally likely to have the prize. It appears the door you chose has a 50/50 chance. Most stick with their initial choice because there is no perceived reason to change.

Time to shatter this illusion with the truth! If you switch doors, you double your probability of winning!

What!?

Solving the problem

When Marilyn vos Savant was asked this question in her *Parade* magazine column, she correctly answered that you should switch doors to have a 66% chance of winning. Her answer was so unbelievable that she received thousands of incredulous letters from readers, many with Ph. D.s! Paul Erdős, a noted mathematician, was swayed only after observing a computer simulation.

It'll probably be hard for me to illustrate the truth of this solution, right? That is the easy part. I can show you in the short table below. You just need to be able to count to 6!

There are only nine different combinations of choices and outcomes. Therefore, I can show you all of them, and we can calculate the percentage for each strategy.

You Pick	Prize Door	Don't Switch	Switch
1	1	Win	Lose
1	2	Lose	Win
1	3	Lose	Win
2	1	Lose	Win
2	2	Win	Lose
2	3	Lose	Win
3	1	Lose	Win
3	2	Lose	Win
3	3	Win	Lose
		3 Wins (33%)	6 Wins (66%)

Here's how you read the table of outcomes for the Monty Hall problem. Each row shows a different combination of initial door choice, where the prize is located, and the outcomes for when you "Don't Switch" and "Switch." Keep in mind that if your initial choice is incorrect, Monty will open the remaining door that does not have the prize.

The first row shows the scenario where you pick door 1 initially, and the prize is behind door 1. Because neither closed door has the prize, Monty is free to open either, and the result is the same. In this scenario, if you switch, you lose, or if you stick with your original choice, you win.

For the second row, you pick door 1, and the prize is behind door 2. Monty can only open door 3 because otherwise he reveals the prize behind door 2. If you switch from door 1 to door 2, you win. If you stay with door 1, you lose.

The table shows all the potential situations. We just need to count the number of wins for each door strategy. The final row shows the total wins and confirms that you win twice as often when you accept Monty's offer to switch doors.

I hope this empirical illustration convinces you that the probability of winning doubles when you switch doors. The challenging part is to understand *why* this happens! To understand the solution, you must first know why your brain screams the wrong solution: 50/50. Our brains are using an incorrect assumption for this problem, so we can't trust our answer.

Typically, we think of probabilities for independent, random events. Flipping a coin is a good example. The likelihood of heads is 0.5, and we obtain that simply by dividing the specific outcome by the total number of outcomes. That's why it *feels* so right that the final two doors each have a probability of 0.5.

For this method to produce the correct answer, the process you are studying must be random. However, Monty acts with intention rather than randomness.

The only random portion of the process is your first choice. When you pick one of the three doors, you have a 0.33 probability of selecting the correct one. The "Don't Switch" column in the table verifies this by showing you'll win 33% of the time if you stick with your initial random choice.

The process stops being random when Monty Hall uses his insider knowledge about the prize's location. It's easiest to understand if you think about it from Monty's point of view. When it's time for him to open a door, there are two doors he can open. If he chose the door using a random process, he'd do something like flip a coin.

However, Monty is constrained because he doesn't want to reveal the prize. He selectively opens only a door that does not contain the prize, which increases the probability that the other door contains the prize. The result is that the door he doesn't show you and lets you switch to is more likely to contain the prize.

Here's how it works.

The probability that your initial door choice is wrong is 0.66. The following sequence is completely deterministic when you choose the wrong door. Therefore, it happens 66% of the time:

1. You pick the incorrect door by random chance. The prize is behind one of the other two doors.
2. Monty knows the prize location. He opens the only door available to him that does not have the prize.
3. By the process of elimination, the prize must be behind the door he does not open.

Because this process occurs 66% of the time and always ends with the prize behind the door that Monty allows you to switch to, the "Switch To" door *must* have the prize 66% of the time. That matches the table!

Another way to think about it is that Monty actually offers you two doors. He opens one and then offers the other. Those two doors collectively have a two-thirds chance of having the prize, and there's no reason to pick the opened one.

The solution to the Monty Hall problem seems weird because our mental assumptions for solving it do not match the actual process. Subconsciously, we based our mental assumptions on independent, random events. However, Monty knows the prize location and uses this knowledge to affect the outcomes non-randomly. The results make sense after understanding how Monty uses his knowledge to pick a door.

I have written about the Monty Hall Problem on my website. And that blog post has received more comments than any other. Readers have filled the comment section with doubts that switching doubles your chances of winning even after reading the explanation for why it is true. Most doubters have *interesting* reasons for why it must be 50/50. Many of these comments are from students and I suspect the Dunning-Kruger Effect is occurring.

So, as you read this, I wouldn't be surprised if you find it hard to accept. Just know that both probability theory and computer simulations that play the game millions of times agree that switching doubles your chances. And remind yourself that Monty isn't acting randomly. He's selectively removing non-prize doors from the game.

There is no doubt among statisticians that the correct answer is to switch.

Let's try another probability puzzle!

Birthday Problem

The Birthday Problem asks, how many people do you need in a group to have a 50% chance that at least two people will share a birthday? Go ahead and think about that for a moment. The answer surprises many people.

Many people guess 183 because that is half of all possible birthdays, which seems intuitive. Unfortunately, intuition works as well for solving this problem as it does for the Monty Hall Problem. So, let's get straight to calculating probabilities for people sharing birthdays.

For these calculations, we'll make a few assumptions. First, we'll disregard leap year. That simplifies the math and doesn't change the results by much. We'll also assume that all birthdays have an equal probability of occurring and are independent events. The people already in the group don't affect the likelihood of the next person's birth date.

Let's start with one person and then add people in one at a time to illustrate how the calculations work. For these calculations, it is easier to calculate the probability that no one shares a birthday. We'll then subtract that probability from one to derive the probability that at least two people share a birthday.

1 – Probability of no match = Probability of at least one match

For the first person, there are no birthdays already covered, which means that there is a 365/365 chance that there is not a shared birthday. That makes sense. We have just one person.

$$1 - \left(\frac{365}{365}\right) = 0$$

Now, let's add in the second person. The first person covers one possible birthday, so the second person has a 364/365 chance of not sharing the same day. We need to multiply the probabilities of the first two people and subtract from one.

$$1 - \left(\frac{365}{365}\right) * \left(\frac{364}{365}\right) = 0.0027$$

For the third person, the previous two people cover two dates. Hence, the third person has a probability of 363/365 for not sharing a birthday.

$$1 - \left(\frac{365}{365}\right) * \left(\frac{364}{365}\right) * \left(\frac{363}{365}\right) = 0.0082$$

Now, you're seeing the pattern for calculating the probability for a given number of people. Here's the general form of the equation:

$$1 - \left(\frac{365}{365}\right) * \left(\frac{364}{365}\right) * \left(\frac{363}{365}\right) * \cdots * \left(\frac{365 - n + 1}{365}\right)$$

I can calculate and graph the probabilities for any size group using Excel. This spreadsheet is available at my website on the resource page for this book. See the URL at the end of this book's Introduction.

Probability of Shared Birthday

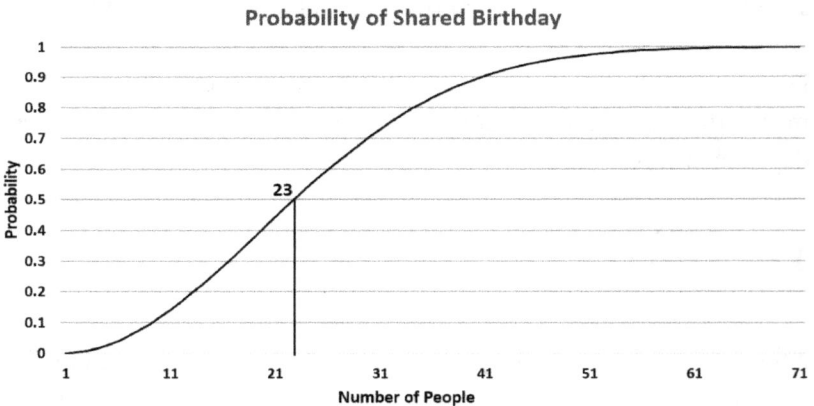

By assessing the probabilities, the answer to the Birthday Problem is that you need a group of 23 people to have a 50.73% chance of people sharing a birthday. Most people don't expect the group to be that small. Also, notice on the chart that a group of 57 has a probability of 0.99. It's virtually guaranteed!

Why is the Group Size So Small for the Birthday Problem?

Like the Monty Hall Problem, most people think the answer of 23 for the Birthday Problem is surprising and hurts their brain a bit!

However, the answer is entirely correct. Let's examine why the answer is counterintuitive.

Often, people think of their birthday and the probability that someone will match that specific date. However, the problem asks about any two individuals sharing a birthday. That means you must compare all possible pairs of individuals. Assessing all pairs causes the number of comparisons to increase rapidly—and therein lies the source of confusion. People tend to think linearly, but the number of pairs increases exponentially.

The formula for the number of comparisons between pairs of N people is: $(N*(N-1))/2$. As you can see in the following table, the number comparisons snowballs to 253 for only 23 people!

People	Pairs
1	0
2	1
3	3
4	6
5	10
6	15
7	21
8	28
9	36
10	45
11	55
12	66
13	78
14	91
15	105
16	120
17	136
18	153
19	171
20	190
21	210
22	231
23	253

For sharing a birthday, each pair has a fixed probability of 0.0027 for matching. That's low for just one pair. However, as the number of pairs increases rapidly, so does the likelihood of a match. With 23 people, you need to compare 253 pairs. With that many comparisons, it becomes difficult for none of the birthday pairs to match.

When there are 57 people, there are 1,596 pairs to compare, and it's virtually guaranteed with a 0.99 probability that at least one pair will match birthdays.

I love problems like this where intuition leads you astray, but math saves the day!

Linear vs. Exponential Thinking

The birthday problem reminds me of a math problem I learned in elementary school that also forces you to think exponentially.

The problem asks whether you would rather receive $10,000 or take a penny on day one and double that amount for 30 days: 0.01 + 0.02 + 0.04 + 0.08, etc. Like everyone else in the class, I took the $10,000. However, the smart choice is to double the penny daily because you'll have accumulated over $10.7 million on the 30th day!

Like the Birthday Problem, there's no trick, just simple math that involves doubling and adding. Our brains think linearly, even in the face of an exponential progression.

There's also a bit of anchoring bias going on. Because the problem mentions $10,000, your mind is thinking in that general region of that dollar amount, making the $10.7 million seem even more extreme.

False Positive Paradox

You might be thinking those previous examples sound artificial. They're designed to mislead and don't necessarily reflect real life. Perhaps. But they effectively show how your intuitive gut feeling can be totally wrong for even relatively simple problems as choosing a door, matching birthdays, and doubling money.

Let's look at a more real-life scenario. What does a positive test result for a medical condition really mean? This paradox is a specific form of cognitive bias known as the base rate fallacy. I'll cover that in the next section, but it involves the difficulty humans have in successfully merging case-specific and base-rate information to find the correct answer. We touched on this issue in the Representativeness Heuristic section in the previous chapter. Here we'll delve into it more deeply.

Can you correctly combine the case specific and base rate information for the following scenario?

Suppose a patient gets a positive test result for a severe condition. Here are the details:

- **Case specific**: The patient has a positive test result using a test that is 95% accurate. 95% of the time, the test produces a positive result when a person has the disease.
- **Base Rate**: 1 in 1,000 people have this condition.

Given this information, what is the probability that the patient with the positive test result has the condition?

The most common answer is 95%, which sounds logical given the test's high accuracy.

However, the correct answer is about 2%. If that's a big surprise, it's because you fell victim to the base rate fallacy! Specifically, you didn't incorporate the base rate of the disease's occurrence in the population, which has a probability of 0.001.

Let's solve this problem. I'll walk you through the calculations incorporating the base rate. Additionally, the resource webpage for this book has a simple Excel worksheet that calculates the answer and allows you to change the parameters.

The trick to avoiding the base rate fallacy is to correctly evaluate the case-specific information within the context of the population's overall probability.

For the medical test example, the test is very accurate for specific cases (95%), but we need to interpret that in a context where the medical condition is rare—a base rate of only 0.001. When the overall likelihood is low, we need to worry about the role of false positives.

To answer the question, consider the following:

We know that in the population, 0.001 have the condition, so $1 - .0001 = 0.999$ don't have it.

Also, the test has a 0.95 true positive rate, meaning it has a $1 - 0.95 = 0.05$ false positive rate.

Now, let's apply that information to a population of 1 million to find the numbers of true positives and false positives.

Cases & True Positives

Let's take our population of 1,000,000 and multiply it by the condition's base rate to find the number of cases: $1,000,000 * 0.001 = 1,000$.

Now, we'll take our 1000 cases and multiply them by the test's accuracy rate to find the number of true positives: $1000 * 0.95 = 950$.

We'd expect to obtain 950 true positives in our population.

Non-Cases & False Positives

Now, we'll find the number of non-cases using its base rate: $1,000,000 * 0.999 = 999,000$.

999,000 don't have the condition, but they can receive false positives when they take the test. Let's use the false positive rate to calculate the number of false positives: 999,000 * 0.05 = 49,950.

Putting It All Together

Most (950) of the 1,000 people with the condition obtained true positive test results. That's great!

However, for those without the condition, a whopping 49,950 get false positive results! While the false positive rate is low (0.05), so many people don't have the condition that the test produces far more false positives than true positives. That explains why the base rate fallacy gives us an extremely biased idea!

Types of Positive Results	Number
True	950
False	49,950
Total	50,900

Hence, only 950 out of 50,900 total positives are true: 950 / 50,900 = 1.87%.

In the example, our calculations show that the probability of actually having the disease given a positive test result is about 1.87%. The low base rate (1 in 1000) dramatically impacts the likelihood of having the disease with a positive result, even when the test is 95% accurate.

The base rate fallacy causes most people to entirely misjudge the meaning of a positive result for this test due to the rarity of the disease.

Several notes. If you change the population's size, the answer remains the same. There is a more complex solution using Bayesian probabilities, but the results are the same. This approach better emphasizes how the base rate affects the ratio of true to false positives.

Of course, this positive test result interpretation applies specifically to the conditions I defined for the example. Don't apply them generally! In fact, doctors frequently limit testing to high-risk populations to reduce false positives.

Consider a patient who has several risk factors for a disease, making them high risk. If doctors test based on those risk factors, it increases the base rate prevalence to higher values such as 0.01 or 0.1, whereas our example used 0.001. You can use the spreadsheet to see how that changes the results.

This false positive paradox example is a specific instance of the base rate fallacy. Let's look at this cognitive error in a more general sense.

Base Rate Fallacy

The base rate fallacy is a cognitive bias that occurs when a person misjudges an outcome by giving too much weight to case-specific details and overlooks crucial probability information that applies to all cases in a population. That vital probability is the outcome's base rate of occurrence in the population. The previous false positive paradox example is a specific instance of this broader phenomenon.

In essence, people misinterpret outcomes because they get tripped up by specific details and overlook the overall frequency of occurrence. They tend to make predictions using similarity rather than statistical likelihoods.

But why does this happen? It's our brain's preference for narrative over numbers, for compelling details over cold, hard statistics. Let's see an example of that in action.

Imagine you're at a local park and notice someone using a telescope and making detailed notes in a notebook. They're dressed casually but have a badge with astronomical symbols. Based on these specific details, you conclude that this person is an astrophysicist or a professional astronomer. However, this assumption could be an instance of the base rate fallacy.

While the equipment and the badge suggest a rare and specific career in astronomy, the base rate of encountering a professional astronomer in a general population is miniscule. Think of the equipment and badge as a positive test result and the base rate is the number of professional astronomers out of the entire population. False positives outweigh true positives in this context, just like the medical test result example.

It's more likely this person is an amateur astronomy enthusiast or a teacher preparing for a class. By focusing on the specific observable details and ignoring the base rate, your conclusion might be far from the truth.

Why do we fall prey to this fallacy? The answer lies in our cognitive wiring. Our brains love stories and specific details; they engage us emotionally and are easier to recall than abstract probabilities. While often helpful, this preference can lead us to ignore the statistical realities that should guide our decisions.

Let's define "base rate." Imagine it as the backdrop of probability against which we should weigh any new information. For instance, in the false positive paradox example, only 1 in 1000 people have a rare disease, which is the base rate. Simple, right? Yet, when faced with specific, often vivid details, our brains tend to put this crucial statistical context in the backseat.

The base rate fallacy illustrates a common cognitive challenge: integrating specific situational details with broader, more generalized data. People typically rely on general base-rate information when it's the only data available but struggle when both specific and general information are present.

Our minds prefer situational details over statistical probabilities! In the next section, we'll see this issue raise its head again with the conjunction fallacy.

This cognitive quirk can affect our judgments in various contexts, from medical diagnosis to financial forecasting, legal decision-making, and everyday life choices. The base rate fallacy can lead us astray, causing misjudgments and inefficiencies, as it skews our perception of risk and likelihood.

In our social interactions, for instance, the base rate fallacy might cause us to overlook how someone has behaved in similar past scenarios (the base rate), leading us to form judgments based on immediate, observable traits. This approach can oversimplify complex human behavior.

In a financial setting, an investor might make decisions based on recent events rather than long-term trends (the base rate).

In the legal system, the base rate fallacy can significantly influence courtroom decisions. Judges and juries, for example, can overly focus on the compelling details of a specific case or testimony while neglecting broader statistical data, such as crime rates or the likelihood of certain events occurring. This imbalance can lead to misinterpretation of the evidence and potentially unjust verdicts.

Perhaps you've heard that the majority of COVID deaths occurred with people vaccinated against the disease? According to CDC data in 2022, 39% of COVID deaths occurred in the unvaccinated while 61% were vaccinated. If that leads you to think the vaccines are ineffective, welcome to another example of the base rate fallacy!

This fallacy occurs when we neglect the broader context—in this case, the population's vaccination rate. Suppose a large portion of the population is vaccinated. In such a scenario, even an effective vaccine might not prevent every case, resulting in a scenario where vaccinated individuals still make up a substantial portion of COVID-19 cases. However, this doesn't imply vaccine ineffectiveness.

To accurately assess vaccine efficacy, compare the infection rate in vaccinated and unvaccinated individuals. Consider these statistics from September 2022. Among the 203 million vaccinated individuals, there were 7,800 deaths, corresponding to a death rate of approximately 38 per 1 million vaccinated people. Conversely, the unvaccinated population of 55 million had 5,200 deaths, indicating a significantly higher death rate of 95 per 1 million unvaccinated individuals.

These figures starkly illustrate the heightened risk for unvaccinated adults, who are about 2.5 times more likely to die from the virus compared to their vaccinated counterparts. This comparison underscores the vaccine's effectiveness in reducing the mortality rate due to COVID-19, thus highlighting the misinterpretation inherent in the base rate fallacy.

In a later chapter, we'll revisit this COVID vaccination example because accounting for other factors reveals that the vaccine is even more effective than this simple analysis suggests.

In all these examples, you need to correctly balance case-specific and general information to obtain a more accurate understanding of complex scenarios. This approach helps you avoid the base rate fallacy trap.

This fallacy is a pervasive cognitive bias that significantly impacts our decision-making. As we've seen, this fallacy occurs when we give too much weight to specific details of a scenario while overlooking the general probability or base rate of an event happening in a population.

Understanding the base rate fallacy is crucial for making informed, rational decisions. Remember, the next time you face a decision or form an opinion, pause, and think about the base rate. Try to combine the specific and general information correctly.

Conjunction Fallacy

The conjunction fallacy is a cognitive bias that occurs when someone mistakenly believes that two events occurring together are more likely than either of the two events alone. In other words, it's the mistaken belief that a precisely detailed, multifaced outcome is more likely to occur than a more generalized version of that outcome.

This phenomenon occurs when individuals concentrate on how two events are connected rather than evaluating the likelihood of each event independently. This focus on the interplay between events can result in flawed or misguided judgments, as it shifts attention away from considering the actual probabilities of the individual occurrences. It leads us to overvalue the chances of simultaneous events. This misjudgment can leave us ill-prepared for the outcomes, as our expectations don't align with the reality of the probabilities involved.

Here is the classic example of the conjunction fallacy in the literature, known as the Linda Problem.

Linda is 31 years old, single, bright, and outspoken. She majored in philosophy. As a student, she was a philosophy major deeply concerned about discrimination and social justice issues and participated in anti-nuclear demonstrations.

Based on the description, which of the following two statements is more probable?

1. Linda is a bank teller.
2. Linda is a bank teller and is active in the feminist movement.

The correct answer is #1. However, many choose option 2 because it seems more plausible given Linda's background. Specifically, in the study conducted by Tversky and Kahneman where the Linda Problem first appears, about 85% of participants chose this option despite it being statistically less likely than option 1 (Linda being just a bank teller). This striking result highlights the intuitive appeal of detailed narratives over statistical reasoning, a critical aspect of the conjunction fallacy.

Our brains prefer the narrative over statistical probabilities. In this sense, the conjunction fallacy is like the base rate fallacy.

The likelihood of two events occurring in conjunction (Linda being a bank teller and being an active feminist) is always equal to or less than the likelihood of either occurring alone.

Let's see why that must be true.

Imagine a world where every bank teller is a feminist—that's a 100% probability. In this scenario, all bank tellers are also feminists. Hence, Linda's chances of being both a bank teller and a feminist are precisely the same as her chances of being a bank teller because the numbers are equal.

However, in any situation where the probability is not a perfect 100%, not all bank tellers are feminists. In this scenario, the number of feminist bank tellers must be lower than the total number of tellers. Hence, the likelihood of Linda being a bank teller and a feminist (option 2) is less than the probability of Linda being just a bank teller.

Thus, from a statistical viewpoint, it's more likely that Linda is simply a bank teller. The math tells us that adding more specifics (like being an active feminist) reduces the probability, even when those specifics seem to align perfectly with Linda's character.

Let's examine the conjunction fallacy using the following Venn diagram: one circle represents bank tellers and the other represents feminists.

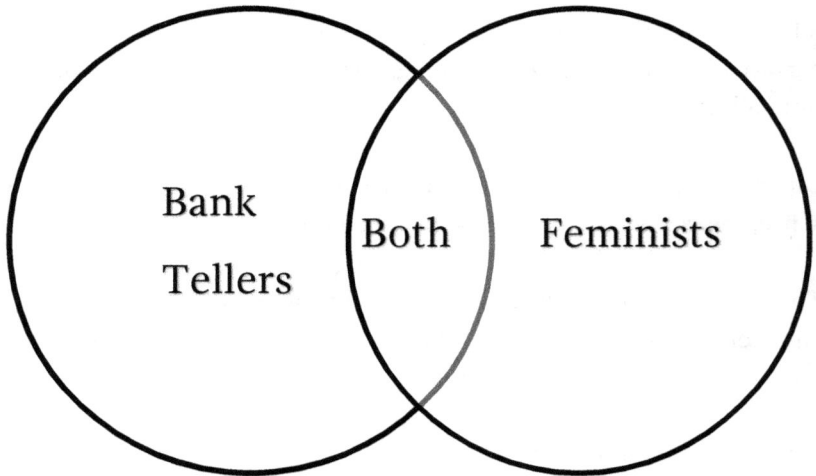

The bank teller circle contains all individuals who are only tellers and those who are both tellers and feminists. That combination is a subset that cannot be larger than the tellers. Even if the two circles overlapped such that all bank tellers are feminists, that number cannot be larger than the number of tellers. That corresponds to the earlier 100% probability scenario.

It's easy to think that a more specific condition is more likely than a broader one because the combination feels right. But that's how the conjunction fallacy trips us up. The Venn diagram shows the truth: the broader condition has more room to happen. It's counterintuitive – the more details you add, the less likely something is, even though it might sound more believable.

Again, our brains prefer a narrative over statistical probabilities.

Tversky and Kahneman suggest that the conjunction fallacy stems from our reliance on the representativeness heuristic we covered earlier. This heuristic involves making judgments based on how much an option resembles our mental image or stereotype. In the case of Linda, option 2 feels more fitting or 'representative' of the detailed portrait painted of her, leading many to choose it. This feeling happens even though, from a purely mathematical standpoint, option 2 is less likely.

This insight sheds light on how our perceptions and biases can override logical probability in decision-making. It's all about how we let specific details skew our judgment.

To avoid the conjunction fallacy, assess the likelihood of each event on its own merits instead of getting caught up in how these events might relate to each other. By methodically evaluating the probability of each occurrence, you can make better assessments and decisions, free from the misleading influence of perceived connections between events.

A comprehensive health survey encompassed a diverse group of adult men from British Columbia, covering various ages and professions. Included in this group was Mr. G., who was randomly chosen from the roster of participants.

Consider which of these scenarios is likelier: (select one)

1. G. has experienced at least one heart attack.
2. G. has experienced at least one heart attack and is above the age of 55.

I bet you fared better on this one!

Because the likelihood of combined events can never surpass that of individual events, logically, the first option is statistically more probable even though we associate heart attacks with older people.

With the conjunction fallacy, people wrongly assume multiple specific conditions are more probable than one. Adding more details to a scenario, while making it seem more plausible, actually decreases its likelihood. This exploration sheds light on the subtle ways our minds work and the importance of critical thinking in everyday decision-making.

Gambler's Fallacy

The gambler's fallacy is a cognitive bias that occurs when people incorrectly believe that previous outcomes influence the likelihood of a random event. It's also known as the "Monte Carlo Fallacy," named after a casino in Monaco, where it was famously observed in 1913.

The fallacy assumes that independent, random events are "due" to balance out over time. Recall the discussion about independent events in probability theory. For this type of event, previous events don't influence the probability of future events.

Have you ever flipped a coin and thought it must be "heads" next because it's been "tails" several times in a row? Or have you been in a casino and believed a particular slot machine is due to pay out because it hasn't hit the jackpot in a while? If so, you've fallen victim to the gambler's fallacy.

If you flip a coin and get ten heads in a row, the probability of the following coin toss being heads again is still 50/50. Independent, random processes don't have a "memory" of previous events.

The gambler's fallacy is not limited to gambling scenarios. It can also occur in situations that involve chance events, such as predicting the gender of newborn babies. Consider a family that has three boys and is expecting their fourth child. The parents might fall victim to the gambler's fallacy by assuming that they are "due" for a girl and that the probability of having a girl has increased since they already have three boys.

However, random processes determine the gender of a baby, and the sex of the previous children does not affect the probability of having a boy or a girl. The chances of having a boy or a girl are 50/50, regardless of the previous outcomes. Therefore, assuming the probability of having a girl has increased exemplifies this cognitive bias.

The gambler's fallacy is also evident in the stock market. Investors tend to sell stocks that have been rising, believing that the sequence of gains means that a decline is due.

The gambler's fallacy can be dangerous in many situations, but it's especially problematic in gambling. Some people believe they can beat the odds by using it to their advantage. They might increase their bets or change their strategy based on the false belief that a particular outcome is "due" to happen. Unfortunately, this can lead to significant financial losses and even addiction.

Let's take an example to illustrate the gambler's fallacy. Imagine you're playing roulette, and the ball has landed on "red" for the last five spins. You might think it's more likely to land on "black" next, but each spin is independent, and the odds of "red" or "black" are always the same. Betting on "black" just because "red" has come up several times in a row is the gambler's fallacy in action.

Researchers started evaluating the gambler's fallacy during the 1960s when studying how the mind processes probabilities.

In early experiments, researchers asked participants to predict which of two colored lights would illuminate next. When chance caused one color to light up several times in a row, participants were likelier to guess the other color as the next one to light up. They wrongly assumed the lights were somehow "due" to switch to the other color, even though the probability was always 50/50.

This experiment and others helped researchers identify the cognitive biases that underlie the gambler's fallacy.

The gambler's fallacy occurs because we look for patterns and meaning in everything we encounter, including the outcomes of random events. When we observe a series of similar events, such as several coin flips resulting in heads, our brains naturally assume that the next event will be tails to "balance it out."

This cognitive bias is driven by the incorrect assumption that random chance is fair and balanced rather than just arbitrary. People tend to believe that if an event occurs too frequently or not frequently enough, it must be due to some external factor or influence. Independent events are entirely random, and no external force balances their outcomes. Past outcomes don't influence future results.

Gambler's fallacy can affect anyone assessing the likelihood of a future event by looking at similar past events. This problem occurs in personal and professional contexts, as our brains naturally try to identify patterns and predict the future. However, errors arise when we apply this approach to independent, random events that are inherently unpredictable. Mistakenly assuming these events follow patterns can significantly impact our predictions and subsequent decision-making processes.

To minimize this problem, be aware of circumstances where you feel something is "due" or will "balance itself out." Carefully evaluate the situation and determine whether it is a random process involving independent events. If so, remind yourself that past events do not affect future outcomes. The next coin toss is always 50/50 regardless of the number of consecutive heads before.

Alright, that seems simple enough. Don't make decisions that require random, independent events to have a memory and adjust to correct a balance.

But wait, there's more! There are several complications surrounding this fallacy. These are snags where you might go out of your way to avoid the gambler's fallacy just to walk smack dab into a different type of problem.

Law of Large Numbers

Misunderstanding the law of large numbers can also cause people to fall for the gambler's fallacy.

The law of large numbers states that as the number of trials increases, sample values tend to converge on the expected result. For example, the more times you flip a coin, the percentage of heads tends to converge on 50%.

Consequently, if you've had only 30% heads, you might think that heads are "due," and their probability is higher. However, that belief is mistaken.

The Law of Large Numbers applies to a long series of independent events, not just to a few or the next one.

In our coin toss example with 30% heads, the law states you'd expect the percentage to be closer to 50% after flipping the coin *many* more times. However, regardless of your location in a sequence of coin tosses, the following coin toss always has a 50% chance of heads.

Independent vs. Dependent Events

Be careful in determining whether you're assessing independent or dependent events. There are several potential minefields to navigate!

Independent events are those in which one event doesn't affect the outcome of the next event.

However, there are *dependent* events where one outcome *does* influence the next. For example, when you draw a card from a standard deck and do not replace it, if the card is NOT an ace, the probability that the next card is an ace is slightly higher.

The gambler's fallacy occurs when you assess independent events as if they were dependent. Conversely, if you model dependent events as though they were independent, you also run into problems.

It might seem easy to tell the difference, but professionals who should've known better have made drastically incorrect forecasts by confusing independence and dependence.

Long-Term Capital Management (LTCM) was a prominent hedge fund in the late 1990s. It used sophisticated financial models and assumed market movements were independent from day to day. However, during the Russian financial crisis of 1998, market conditions showed that movements were not independent, as extreme events were more likely to be followed by more extreme events.

This incorrect assumption of independence led to a massive underestimation of risk, contributing significantly to LTCM's eventual collapse and bailout. LTCM was famous for using quantitative models for trading and even they made such a fundamental error. Its collapse in 1998 was a significant event in financial history.

Similarly, the 2007-2008 housing market crash was partly due to a critical misjudgment in financial risk assessment. Financial institutions believed that mortgage defaults were independent events, underestimating the risk of widespread failures.

However, these defaults were actually interdependent, linked by broader economic factors like housing prices and interest rates. When the market declined, a ripple effect of defaults occurred, revealing that the risks were much more interconnected than previously assumed. This oversight in recognizing the dependencies between financial risks significantly contributed to the severity of the crash, underscoring the need for more complex risk assessment models in finance.

Even seasoned professionals fall victim to these errors!

Closing Thoughts

This chapter examined our abilities to use probabilities to make decisions. I hope the probability problems were fun and pointed out that mathematically correct answers don't always feel right. Yes, our intuition can be way off!

The Monty Hall problem, a classic example, demonstrates the importance of understanding the process of calculating probabilities. It's not just a theoretical puzzle—it has practical implications. Is the process random or not? What is effect do non-random interventions have? Just because you have two choices doesn't mean it's a 50/50 situation. This problem challenges our intuition and shows how probabilities can be counterintuitive in real-life scenarios.

The Birthday Problem, the doubling pennies, and the False Positive Paradox are just math and calculations. No tricks. Our intuition is just wrong about them. And that's another crucial point to keep in mind. We often say, "trust your gut." But that doesn't always work in a data analysis context!

The base rate, conjunction, and gambler's fallacy all show that we make systematic, predictable errors when mentally modeling probabilities.

When you hear that about two-thirds of COVID fatalities as of 2022 were vaccinated, it sure makes it seem like the vaccines aren't working. But that wasn't the case at all. It becomes evident when you consider the base vaccination rate (most people were vaccinated) and compare the death rates between the vaccinated and unvaccinated.

Even professional analysts at a brokerage famous for its modeling majorly messed up!

Probabilities are tough. You need to assess each instance carefully and deeply understand what is happening. Some common issues trip people up, but there isn't a single magic bullet other than carefully considering the problem's specifics and then using the correct methods.

Be sure to understand the true nature of your subject matter!

Statisticians and data analysts have developed various methods and tools to address the cognitive shortcomings in the first two chapters. In the following chapters, we shift gears and focus on them. However, we're not entirely leaving cognitive biases behind because they continue to play a role in how we apply analyses and interpret their results. After all, the entire process rests on the cognitive foundation of the Thinking Analytically pyramid. Understanding and mitigating these biases is crucial for making informed decisions.

CHAPTER 3

Data Quality: Sampling

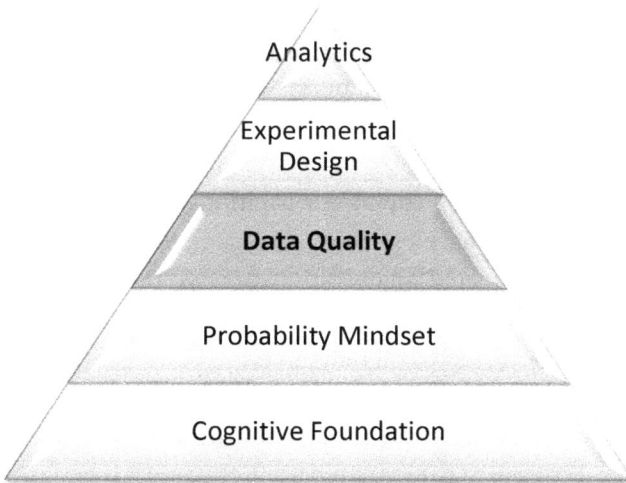

In these times where data drives decisions, the ability to dissect and comprehend the nuances of data is more crucial than ever. The third level of the thinking analytically pyramid involves being a data detective, a pivotal stage in our journey towards analytical mastery. You must understand how the data were collected and measured because that affects the conclusions you can draw from them.

Data quality is a lens through which we must view all analyses. This examination tells us how much we should trust the results. This level of the Thinking Analytically pyramid incorporates the idea of garbage in and garbage out (GIGO). Even the most sophisticated analysis will produce garbage results if you start with poor-quality data.

As we dive into the Data Quality pyramid level, we will unravel the intricacies of sampling methods and measurement nuances, two pillars upholding accurate data interpretation. Remember, the power of data lies not just in quantity but in quality.

Understanding data is akin to assembling a complex puzzle. Every piece—how it's gathered, the context of its origin, and its measurement method—plays a vital role in the larger picture. This chapter and the next are about processes and procedures and how they can affect the numbers.

- How were the participants selected?
- How were the measurements performed?

While these topics might sound less exciting, they are crucial for producing trustworthy analyses. You'll see examples of how problems in these areas have created analytical errors. By the end of these two chapters, you will be adept at questioning data, equipping you with the critical eye of a seasoned data detective.

Many data analysts and those who review or use the results overlook these steps. It's easy to get wrapped up in the analysis and its interpretation. We all want to see results. What's the answer?!

My goal isn't to teach you everything there is to know about sampling methods and measurement systems assessment—each of those are large subject areas—but you need to know how to question the data and look for problems. Even if you aren't conducting the analysis yourself, this information will help you evaluate data others have analyzed.

For analysts in big data who don't control how their data are collected, understanding the differences between the best methods and how your data were gathered provides valuable insights to potential biases.

In this chapter, we examine how sampling methods affect data quality. Researchers use these procedures to obtain the people or items they'll measure in their study. You'll also learn about various problems that can degrade an otherwise good sample. Collectively, these methods and related issues can profoundly affect the results. We'll focus on sampling precision and bias. This chapter addresses bias more thoroughly, and Chapter 7 expands on sampling precision.

The following chapter looks at data quality in terms of measurement error.

Descriptive Statistics: The Simplest Case

Before we get to sampling methods, it's essential to understand the simplest case, where you measure a small group and apply the results only to that group. Statisticians refer to that process as descriptive statistics.

Descriptive statistics describe a sample. That's pretty straightforward. You simply take a group that interests you, record data about the group members, and then use summary statistics and graphs to present the group properties. With descriptive statistics, there is no sampling uncertainty because you describe only the people or items that you actually measure. You're not trying to infer properties about a larger population.

For example, you want to measure the heights of 30 students in a specific class and break them down by gender. So, you measure the heights and record their genders. Later, you report on the mean heights by gender.

That process is simple and fantastic for understanding that specific class—but nothing else. Your results apply to that class only. You can't draw conclusions about other groups, not even other classes.

Now, you take that same class and divide them into a treatment and control group. Then, you give the treatment group a growth medication. Later, you measure the heights of both groups and find that the treatment group is taller than the control group.

Unfortunately, you can only state the results for that specific group of 30 students. That's not useful scientifically because you have no idea how the drug performs in a broader population. To generalize to a population, you must use inferential statistics.

Inferential Statistics: More Useful but Complicated

Inferential statistics takes data from a sample and makes inferences about the larger population from which the researchers drew the sample. Because inferential statistics aim to draw conclusions from a sample and generalize them to a population, we need to have confidence that our sample accurately reflects the population. This requirement affects our process. At a broad level, we must do the following:

1. Define the population we are studying.
2. Draw a representative sample from that population.
3. Use analyses that incorporate the sampling error.

We don't get to pick a convenient group. Instead, we need to use a sampling method that gives us confidence that the sample represents the population. Unfortunately, gathering a genuinely representative sample can be a complicated process.

Representative Samples

Researchers usually want to learn about a population. After all, if you're studying opinions, attitudes, characteristics, or the effects of a new medication, generalizing the results to an entire population is much more valuable than understanding only the relatively few participants in the study. Think back to the class example in the descriptive statistics section. Knowing how the medication works for only that one class is not helpful!

Unfortunately, populations are usually too large to measure fully. Consequently, researchers must learn about them using a manageable subset. This point is where representative samples come in.

A representative sample is one where the individuals in the sample reflect the properties of an entire population. Consequently, by studying a representative sample, you can estimate the properties of the population from which it was drawn. They're essential for generalizing the results to a population.

How do you tell if a sample represents a population?

Each study needs to define the target population it wants the sample to represent. Researchers must conduct preliminary research to understand this population. Throughout the study, they gather insights about the individuals within the target group.

A representative sample proportionally reflects the attributes of the target population. The demographic characteristics of the individuals in the sample must be similar to those in the population: gender, rural, urban, religion, marital status, income levels, etc. However, the relevant attributes depend on your study area. For example, if it's a health study, you might need to consider other aspects such as health habits, BMI, blood pressure, etc.

A representative sample enables generalization of the results. Without this, the findings apply only to the specific sample studied. If your sample does not resemble the population you are studying, you can't trust that the sample results will apply to the population.

For all these reasons, researchers strongly prefer obtaining representative samples whenever possible, even though this type of sample can be the hardest to get.

Examples

Suppose you are assessing the approval of a controversial new law in a state. You define your population as all adults in the state. Unfortunately, it's impractical to contact all adults. Instead, you need to obtain a representative sample.

Your sample must contain individuals who resemble the whole population by including all demographic groups and have them in the same proportions as the entire population. For example, it's not representative if you have too many rural participants, males, etc.

After collecting your sample, you can administer the survey. The proportion of your sample that approves of the opinion is an unbiased estimate of the population proportion.

Other examples of using representative samples include the following:

- Election polling for a particular jurisdiction.
- Surveys of a specific profession, such as medical doctors.
- Literature preferences of Master's level English students.
- Income distribution among farmers in a specific country.
- Vaccine effectiveness in healthy adults.

Notice how all these examples specify what you're measuring and define a population. To obtain a representative sample, the researchers must first clearly define the population, stating who the researchers are learning about.

In other words, what population should the sample represent? You can only have a representative sample if you know which population it should look like! Additionally, when you generalize the sample results, you need to identify the population about which you are inferring the properties.

Researchers can define the population to meet the needs of their study. For example, I once read an article for a study that defined its population as adult Swedish women (with specific age requirements for inclusion) who have osteoporosis but are otherwise healthy.

After defining your population and drawing a representative sample, you can measure your variables of interest and then generalize them to the population.

Evaluating Representativeness

A representative sample mirrors the population from which it was drawn. How do we evaluate that? You'll first need to understand the difference between sample statistics and population parameters.

Sample statistics are numbers that describe the properties of samples, while parameters are numbers that describe the properties of entire populations.

Both are summary values that describe a group, and there's a handy mnemonic device for remembering which group each describes. Just focus on their first letters:

- Parameter = Population
- Statistic = Sample

For example, the average income for the United States is a population parameter. Conversely, the average income for a sample drawn from the U.S. is a sample statistic. Both values represent the mean income, but one is a parameter and the other a statistic.

Ideally, you use sample statistics to estimate the values of the parameters. Hence, you want to use a suitable method to obtain your sample. If you use the wrong method, you can't use the statistics to estimate the parameters.

While parameters and statistics have the same types of summary values, statisticians denote them differently. Typically, we use Greek and upper-case Latin letters to signify parameters and lower-case Latin letters to represent statistics.

Summary Value	Parameter	Statistic
Mean	μ or Mu	$\bar{x}$ or x-bar
Standard deviation	σ or Sigma	s
Correlation	ρ or rho	r
Proportion	P	$\hat{p}$ or p-hat

Sample statistics are estimates of the relationships and effects in the population. For example, the difference between the sample means of the control and treatment groups is the sample's estimate of the population's treatment effect.

Unfortunately, all sample estimates are wrong to some degree. Sampling error is the difference between the sample statistics and the population parameters. All samples have error because they never exactly represent a whole population. That's challenging, but we must live with it when working with samples.

Unfortunately, we never know the actual population value. So, we never know the true amount of sampling error. However, statisticians have devised a framework for assessing it and identifying potential sources. Indeed, we can estimate some forms of error and use that to test hypotheses. However, more problematic forms of error also exist.

Statisticians divide the types of sampling error into two broad categories:

- Random errors are random differences between the sample and the population. Even with the most carefully gathered sample, they exist, creating a margin of uncertainty.
- Sampling biases are systematic errors that are more problematic.

In the following sections, we'll unravel random sampling error and bias, learn to spot them, understand their impact, and, most importantly, discover strategies to mitigate their influence. Then, we'll move on to the different ways to draw a sample and their pros and cons. This knowledge helps you understand your data better and enables you to become a more discerning interpreter of analyses others have performed.

Sampling Precision and Bias

Sampling error and bias are often unseen yet profoundly influential. These concepts represent the gap between the properties of the population and those of the sample. Understanding them allows us to acknowledge and navigate data imperfections.

We'll explore these characteristics using sampling distributions.

A key concept of inferential statistics is that the sample a researcher draws is only one of an infinite number of samples they could have drawn. Imagine we repeat a study many times. We collect many random samples from the same population and calculate each sample's mean. Then, we graph the distribution of the sample means, which statisticians call a sampling distribution.

In Chapter 7, I cover sampling distributions in greater depth. For now, we'll use them to understand sampling bias and precision.

The preferred condition for a study is that the sampling error should be unbiased and small. This state indicates that repeated samples tend to center on the population value and cluster tightly around it.

Unbiased

Unbiased sampling error tends to be right on target. For instance, representative samples will randomly fluctuate around the correct population values. While the sample estimates won't be exactly right, they should not be systematically too high or low. The average or expected value of multiple attempts should equal the population value. Statisticians refer to this property of being correct on average as unbiased.

In the graph below, the population value is the target that the distribution for representative samples centers on, making them unbiased. Remember that the curves represent sample means from an infinite number of studies rather than individual measurements.

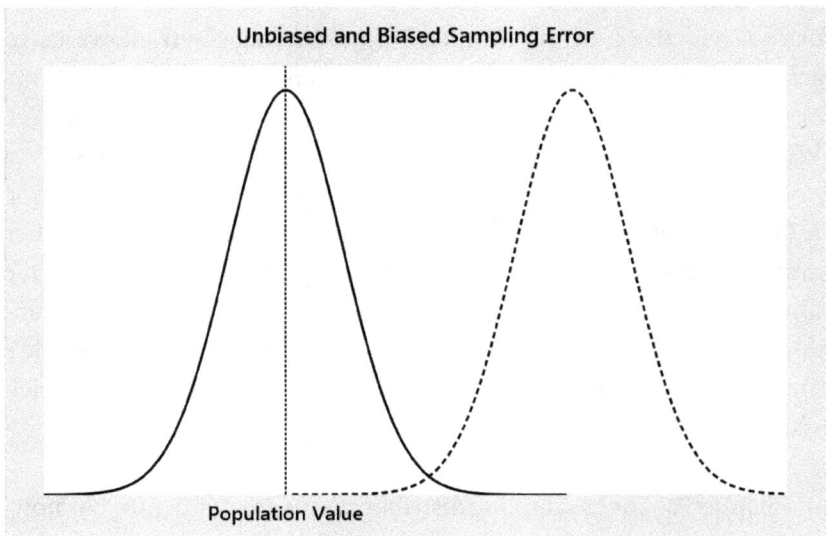

Unbiased and Biased Sampling Error

Population Value

The curve on the right centers on a value that is too high. The methodology behind this study tends to overestimate the population parameter, which is a positive bias. It is not correct on average. Statisticians refer to this problem as sampling bias.

However, the left-hand curve centers on the correct value. That study's procedures yield sample statistics that are correct on average—it's unbiased. The expected value is the actual population value.

Precision

Recognizing that sample statistics are always incorrect to some degree, you want to minimize the difference between the estimate and the population parameter. Large differences are bad!

Precision assesses how close you can expect your estimate to be to the correct population value. When your study has low random sampling error, it produces precise estimates that you can confidently expect to be close to the population value. That's a better position than having high amounts of error that produce imprecise estimates. In that scenario, you know your estimate can be wrong by a significant amount!

For example, you're studying the treatment effect of a new medication. Your estimate is the sample difference between the control and treatment group means. High precision indicates that your sample estimate, while wrong to some degree, is likely close to the population value. Conversely, low precision suggests it can be far away from it.

Sampling distributions represent sampling error and precision using the width of the curves. Tighter distributions represent lower random error and more precise estimates because they cluster more tightly around the population value. Conversely, broad distributions indicate lower precision because estimates tend to fall further from the correct value.

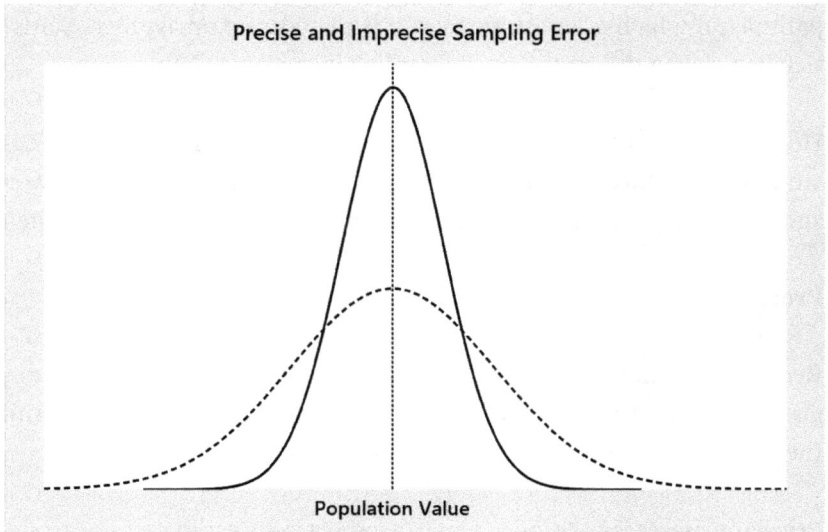

Precise and Imprecise Sampling Error

Population Value

In the graph, both curves center on the correct population value, indicating they're unbiased. That's good. However, the dashed curve is broader than the solid curve because it has more random error. Its estimates tend to fall further from the population value than the solid curve. That's not good. We want our estimates to be close to the actual population value.

Relatively precise estimates cluster more tightly around the parameter value, which you can see in the narrower curve.

Let's start by breaking down the sources of random sampling error and sampling bias and how to manage them.

Reducing Random Error to Increase Precision

Random sampling error produces imprecision. Unfortunately, this error is an inherent consideration when using samples to some degree. Even when researchers conduct their study perfectly, they can't avoid some random sampling error.

Why not?

Randomness alone guarantees that your sample cannot be 100% representative of the population. Chance inevitably causes some error because the probability of obtaining just the right sample that perfectly matches the population value is practically zero. Additionally, samples can never perfectly depict the population with all its nuances because it is not the entire population. Samples are typically a tiny percentage of the whole population.

The only way to prevent random sampling error is to measure the entire population. Barring that approach, researchers can take steps to understand and minimize it.

Given the inevitability of some random error for most studies, the question becomes, how close are the sample estimates likely to be to the correct population values? Or, how good is the precision? The best studies have high precision, where their estimates are close to the correct values.

Random sampling error refers to chance differences between a random sample and the population. This type of error affects an estimate's precision and do not include the biases that I discuss in the next section. Importantly, random errors do not necessarily suggest any problems exist with the study's design or methods. The best studies all have random error but have minimized it.

Two key factors affect random sampling error—population variability and sample size.

- Low variability in the population reduces the amount of random sampling error, increasing the precision of the estimates.
- Larger sample sizes reduce random sampling error, producing more precise estimates.

Of these two factors, researchers usually have less control over the variability because it is an inherent property of the population. However, they can collect larger sample sizes. Consequently, increasing the sample size is the critical method for reducing random sampling error.

Statisticians prefer random sampling error over sampling bias for several reasons. After covering sampling bias, we'll get to those reasons.

Sampling Bias

Sampling bias occurs when sample attributes are *systematically* different from the actual population values, making them nonrepresentative samples. As you saw in the previous graphs, bias exists when the curves center on the incorrect values. These aren't the random errors from before. They are a consistent factor that pushes the entire curve away from the correct population value. Hence, sampling bias produces a distorted view of the population.

Sampling bias often exists when the study is more likely to select specific subgroups or people with particular attributes than others. Later in this chapter, we'll go over bias sources. A common theme behind them is that they disproportionately affect data collection from specific subgroups. This concept often involves human subjects, but it can also apply to samples of objects and animals.

In everyday language, "bias" has a negative connotation. However, statistical bias indicates a systematic tendency for a sample statistic to over or underestimate a population parameter. Challenges in collecting representative samples and design oversights often cause sampling bias rather than intentional deception. In contrast to random error, sampling bias suggests that problems exist in the study's design or methods.

When you don't understand the nature and degree of the sampling bias, it limits generalizations from the sample to the population.

Analysts can generalize findings only to populations that are like their sample.

The leading causes of bias relate to the study's procedures. There are no statistical measures that assess bias using the sample data alone. A sample's properties cannot tell you if it is biased. Instead, you must examine the nuts and bolts of a study's methods and procedures to determine whether they will likely introduce bias.

How does a study select its subjects? Are particular subgroups more or less likely to participate?

Below are some of the top causes of sampling error bias:

- The study did not use a representative sampling method. Hence, the study obtains a sample that misrepresents the population.
- The study sampled the wrong population. Although it might have used a representative sampling method, it somehow drew a sample from the incorrect population.
- Measurement bias. For various reasons, the measurements systematically differ from the correct attributes. Even if the sample is representative, the measurements might not be.
- The study might have started with a representative sample, but one of many bias sources could have affected it. We'll get to them later in this chapter.

Sampling methods are a common source of bias, and we'll examine them and other sources shortly.

Is Random Error or Sampling Bias Worse?

While both sampling error and bias present challenges in statistical analysis, statisticians typically prefer dealing with random error over bias. This preference stems from the inherent nature of these two phenomena.

Random error, a product of natural variation, is an uncertainty that standard statistical techniques can quantify using the sample itself. Additionally, analysts can reduce it substantially by using larger sample sizes. It's a predictable companion in the data analysis journey that can be managed and mitigated with the right tools.

Bias, on the other hand, is more insidious. As a systematic deviation from the truth, bias represents a fundamental flaw in the data collection process, distorting results in a way that's often difficult to detect and correct. Indeed, analysts can't use the sample data alone to detect bias. It misleads interpretation and can invalidate the findings, no matter the sample size.

While random error is a hurdle analysts can manage with the proper tools, bias can lead even the most diligent researchers astray. Therefore, recognizing and minimizing bias is crucial when pursuing genuine insights from data.

Finally, I want to emphasize a critical difference between random error and bias regarding sample size. A larger sample size reduces sampling error and, thereby, increases precision. Think back to the graph with the wide and narrow curves. Increasing the sample size narrows that curve.

Unfortunately, increasing the sample size won't fix bias. It might narrow the curve, but it will still center on the wrong value. Your result will be more precise, but it'll be precisely *wrong*. Consider a loaded die that shows a 6 more often than it should. If you increase the sample size by rolling the die more times, all you'll get is more 6s! Increasing the sample size just allows existing biases to manifest more.

As you'll see later in this chapter, sometimes you start with an unbiased, representative sample, but various bias sources end up skewing the results by the end.

Now that you understand random error and sampling bias in samples, let's see how that applies to drawing samples.

What Are Sampling Methods?

Sampling methods are how you draw a sample from a population.

A population is the complete set of individuals that you're studying. A sample is the subset of the population you measure, test, or evaluate and base your results on. Sampling methods are how you obtain your sample.

In research and inferential statistics, sampling methods are a vital issue. How you draw your sample affects how much you can trust the results! If your sample doesn't reflect the population, your results might not be valid.

In this section, you'll learn more about sampling methods, which produce representative samples, and the pros and cons of each procedure.

Probability vs Non-Probability Sampling Methods

Sampling methods have the following two broad categories:

- **Probability sampling**: Entails random selection and typically, but not always, requires a complete list of the population.
- **Non-probability sampling**: Does not use random selection but some other process, such as convenience. It usually only samples from part of the population.

Probability sampling is typically more difficult and costly to implement. However, in exchange, these processes produce representative samples that researchers can use to make valid conclusions about the population.

On the other hand, non-probability sampling methods are often easier and less expensive, but the trade-off is that your conclusions are questionable. You might not be able to trust the results if you attempt to make inferences about a population.

Probability sampling methods are the gold standard. Even when a study doesn't use this approach, comparing its method against random sampling is valuable to understand how it stacks up. For this chapter, I'll focus on several standard probability methods and then briefly cover some non-probability methods. My goal is for you to learn enough about the strengths and weaknesses of the various methods so you can evaluate a study's data and results.

I also see an interesting analogy between cognitive biases and sampling methods. Some methods correct for these types of issues while others don't.

Recall from Chapter 1 that the information we remember quickly is a subset (i.e., a sample) of all the information we know. It's a biased sample because we're more likely to recall information we agree with and are emotionally charged, recent, and unusual. That's what comes to mind quickly and easily. It's biased and distorts our perceptions.

Non-probability samples are similar, quick, and easy but biased. Cognitive errors and non-probability sampling both emphasize convenience, efficiency, and speed over accurate samples.

Conversely, probability methods intentionally attempt to avoid bias by proportionately sampling all subgroups, including the more difficult-to-find population members. These approaches require more time and resources to accomplish that.

As with overcoming cognitive biases, probability sampling methods produce better, unbiased results by requiring researchers to slow down and exert more effort to obtain a wide-ranging sample.

Probability Sampling Methods

Given the benefits of using representative samples, researchers use a probability sampling method whenever possible. Let's go over the standard methods. They each have pros and cons.

To use a probability method, you'll first need to define the population and develop a sampling frame, which lists all members of your target population.

A critical point to remember about these methods is that they produce representative samples for a *specific* population. Researchers must define that population and create the sample frame. The results apply to that population alone.

For instance, imagine there are two newspapers and they each obtain a representative sample from a voting district. One newspaper's district leans conservative, while the other's is more liberal.

Now, imagine the two newspapers ask their samples the same survey questions about political issues. You'd expect the results for the two surveys to differ greatly—probably diametrically opposed on some issues even though both surveys are legitimately representative samples. Both sets of results are correct, but for different populations.

Always interpret the results of inferential statistics in the context of a specific population. As you'll see later in this chapter, it's risky applying them to other groups.

Simple Random Sampling (SRS)

In simple random sampling (SRS), researchers take a complete list of the population and randomly select participants from it. All population members have an equal likelihood of being selected. Out of all sampling methods, statisticians consider this one to be the gold standard for producing representative samples. It's entirely random, leaving little room for accidentally biasing the results.

For example, if you randomly select 1000 people from a town with 100,000 residents, each person has a probability of 1000/100000 = 0.01. That's a simple calculation requiring no additional knowledge about the population's composition. Hence, simple random sampling.

SRS is a probability sampling method that helps ensure the sample mirrors the population. The process proportionately samples from larger subpopulations more frequently than smaller subpopulations.

Suppose the town contains subpopulation A with 40,000 people and subpopulation B with 10,000. Using SRS with a probability of 0.01, the process will tend to enlist 400 from subpopulation A and 100 from B. Hence, the process tends to produce a proportionate representation in the sample that reflects the entire population. You don't even need to know the details about the subpopulations for this process to work!

Simple random sampling requires having a sampling frame that lists all population members and the ability to contact and involve them in your study.

To perform simple random sampling, do the following:

1. Define the population.
2. Create a list of all population members.
3. Assign random numbers to each member.
4. Use a random number generator to select participants until you reach your target sample size.

Alternatively, if the population is a manageable size, you can use a lottery system to draw the sample. Place all the names in a hat and randomly draw your sample. Researchers typically use computers to select participants randomly from a database for large populations.

Procedurally, SRS is the simplest method for obtaining an unbiased sample. While the researchers need a list of the entire population, they don't need other information about that population, its subpopulations, and its features.

Conversely, more complex forms of sampling require researchers to understand the population's characteristics. Then, using that knowledge and a lot of preplanning, they divide the population into strata or clusters and perform other procedures before sampling. With SRS, you just randomly draw from the list until you have enough subjects.

However, simple random sampling has some drawbacks.

First and foremost, this method can be unwieldy and require abundant resources. For one thing, it requires a list of all population members, which can be a tremendous hurdle by itself. Attempting to perform SRS with an incomplete population list causes undercoverage bias and a nonrepresentative sample.

Furthermore, while random selection is beneficial, it also ensures that the subjects are maximally dispersed, making them harder to contact.

SRS can exclude smaller but crucial subpopulations purely by chance. Additionally, this approach produces less precise estimates for subgroups and the differences between subgroups than other probability sampling methods.

While SRS produces the most precise estimates for the entire sample, other probability sampling methods can ensure sufficient numbers from small subgroups to produce a clear picture and increase the ability to compare subgroups.

For example, in our town with 100,000 residents, imagine we're particularly interested in surveying people at least 90 years old. You plan to obtain a sample size of 1000, which is 1 out of 100 residents. However, only 50 people in town are older than 90. Your sample might not include anyone in this vital group! If it does, it'll be a tiny number that doesn't provide a clear picture of this subgroup.

Example

Imagine we are studying a town with 100,000 residents. We want to perform simple random sampling to obtain a sample size of 1000. First, we need to define the population. We'll define it as town residents who pay township taxes and are at least 18 years old.

Next, we must create a complete list of residents meeting those criteria. We'll work with the township tax office to make the list and add all eligible residents.

Finally, we need to select participants randomly from the list. We can use a computer program to do that. Alternatively, we can print out names on slips of paper and draw them from a basket. We keep drawing from the list until we have 1000 names.

Systematic Sampling

Systematic sampling is like SRS but attempts to ease some of the difficulties for researchers. There are several versions of this method.

One form uses a complete list of the population. The researchers randomly select the first subject and then move down the list, choosing every Xth subject rather than using a randomized technique.

The other form does not use a complete list of the population. This sampling method is suitable for populations that are tough to document, such as people experiencing homelessness, because a comprehensive list won't exist. The essential requirement for this sampling method is knowing how to locate them. While imperfect, it's feasible when you can't obtain the complete list.

Suppose you want to survey theater patrons but lack a complete list. Instead, you can use systematic sampling to recruit every 20th person exiting the theater. This approach works because they leave randomly.

This sampling method has some disadvantages. The form that uses a complete list of the population can closely mirror the results of simple random sampling. However, the non-randomness increases the potential for manipulation, even if accidentally. Additionally, patterns in the list can unintentionally create a nonrepresentative sample.

The form that doesn't use a list has more potential problems. Namely, it increases the potential for missing subgroups and acquiring a non-representative sample. This sampling method increases the knowledge you must have about the population and their habits. Without that knowledge, you won't be able to find subjects that reflect the whole population.

Stratified Sampling

In stratified sampling, researchers divide a population into similar subpopulations (strata) and then randomly sample from each stratum.

This sampling method can guarantee the presence of small but vital subpopulations in the sample. Relative to SRS, this method can increase the precision of subgroup estimates and the differences between subgroups but the overall precision is lower. In short, it helps researchers gain a better understanding of the subgroups. Additionally, dividing the whole population into smaller, similar subsets can reduce costs and simplify data collection. Researchers can use approaches they optimize for each stratum rather than the one-size-fits-all approach of SRS.

The drawback is that this sampling method requires additional up-front knowledge and planning. The researchers must know enough about the subgroups to devise an effective strata scheme. Then, they must have sufficient information about all population members to assign them to the correct strata.

Cluster Sampling

Like stratified sampling, the cluster sampling method divides the whole population into smaller groups. However, unlike strata, each cluster mirrors the full diversity present in the population. Then, the researchers draw random samples from some of these clusters.

The primary benefit of this sampling method is that it reduces the costs of studying large, geographically dispersed populations. Using this method, researchers don't need to sample the entire geographic region, only certain areas, because they know individual clusters are similar to the population. Additionally, they don't need to develop a list of potential subjects for clusters they're not sampling. These considerations can significantly reduce planning, administrative, and travel costs.

When researchers can't create a list of the entire population, cluster sampling can be an excellent choice.

On the downside, cluster sampling increases the design complexity. Researchers must understand how well each cluster approximates the whole population. If the clusters don't fully represent the population, results can be biased. In real-world studies, clusters tend to be naturally occurring groups that don't mirror the population, which reduces the ability to draw valid conclusions.

Non-Probability Sampling Methods

Non-probability sampling methods don't use random selection, and they typically don't use a complete sampling frame. While these methods are easier and less expensive, your results are more likely biased, reducing your ability to make sound conclusions.

Researchers often use non-probability sampling methods for pilot studies, exploratory research, and qualitative research. These sampling methods provide quick and rough assessments, help work out kinks in measurement instruments and procedures, and refine the design for a more rigorous study in the future.

Below are several standard non-probability sampling methods:

- **Convenience sampling**: The main criteria for recruiting subjects are those who are easy to contact and willing to participate. There are no inclusion requirements. Online polls are a type of convenience sampling.
- **Quota Sampling**: A study uses a non-random selection of subjects from population subgroups that the researchers define.
- **Purposive sampling**: Investigators use subject-area knowledge to handpick a sample they think will help their study.
- **Snowball sampling**: Researchers use subjects to find and recruit other subjects. This method is helpful when a population is challenging to contact. When recruits help you find more recruits, and those help find even more, and so on, the total number snowballs.

Convenience Sampling

Convenience sampling is a common non-probability sampling method in which researchers use subjects who are easy to contact and obtain their participation. Researchers find participants in the most accessible places and impose no inclusion requirements. Convenience sampling is also known as opportunity or availability sampling.

Examples of convenience sampling include online and social media surveys, asking acquaintances, and surveying people in a mall, on the street, and in other crowded locations.

While the subjects are easy to access, the researchers are unlikely to obtain a sample representing the population accurately. Sampling bias is likely to be high. You cannot generalize the sample results to a population. Sometimes, you might not be fully aware of the populations you're sampling. Who's answering your online surveys? In short, the results you obtain using this approach apply only to your sample.

Convenience samples serve as a sharp contrast to representative samples in terms of being able to generalize the results.

Statisticians rarely recommend this method because being unable to generalize your results beyond the sample is a huge limitation. Despite this weakness, there are a few situations where this method is warranted.

Convenience sampling is most useful for pilot testing. Use it when you're testing your survey instrument and other research protocols. It's an inexpensive way to solve many problems with your study before committing more resources to obtain a representative sample.

This method can also provide initial ballpark estimates in the exploratory stages of research. For example, a company might want quick feedback about new logo candidates and obtain a quick sample. At the very least, it expands the feedback beyond company employees.

In other cases, this approach might be the only viable approach. Researchers might have insufficient time and resources to use a representative sampling method. Consequently, student projects often use convenience samples for this reason. In these cases, the preliminary results can serve as a call for more rigorous studies in the area.

Example: Pepsi Challenge

A classic example of convenience sampling is the Pepsi Challenge, a blind taste test conducted at shopping malls, stores, and other public venues. Participants tasted unmarked cups containing Coca-Cola and Pepsi and then indicated their preference.

The Pepsi Challenge has all the hallmarks of convenience sampling, including using crowded areas to facilitate the easy acquisition of participants and the lack of participation requirements.

As you can see, there are many sampling methods, each with benefits and disadvantages. When designing a study, evaluate the nature of your target population, research goals, and the available time and resources to choose your sampling method.

Obtaining an unbiased representative sample requires satisfying quite a few details. I've highlighted some of them so you can better evaluate the data and understand its limitations.

These concepts aren't just theoretical. Let's examine an actual study in which the sampling method produced bias that distorted the results.

Example: Early Hormone Replacement Studies

Initial research on Hormone Replacement Therapy (HRT) in the 1980s and early 1990s suggested that women undergoing this treatment experienced significant cardiovascular benefits. Unfortunately, these early studies had a critical flaw—a biased sample that systematically misrepresented the general population.

These initial findings primarily stemmed from studies that used a sample of middle-aged, white, higher socioeconomic status, and health-conscious women. These women were often non-smokers, had better access to healthcare, and were generally more vigilant about their overall health. This sample led researchers to believe that HRT was safe and beneficial in reducing the risk of heart disease.

The participants already had a lower risk of cardiovascular issues due to their healthier lifestyle choices and better access to medical care. Unfortunately, the early studies were unaware of this sampling bias and incorrectly attributed the heart health outcomes to HRT rather than the biased nature of the sample. These results should never have been generalized to the general population because the sample misrepresented the general population.

Fortunately, science tends to be a self-correcting field. Scientists review studies, discuss them, and propose improvements.

To address these concerns, the Women's Health Initiative (WHI) launched in 1991 and conducted a large-scale trial that included a more diverse group of women. This study involved women from various backgrounds, ages, ethnicities, and health statuses, aiming to provide a more comprehensive understanding of HRT's effects. The WHI trial's design eliminated many biases present in the earlier studies.

The results of the WHI trial were strikingly different from the earlier studies. The WHI found that HRT did not confer the cardiovascular benefits previously reported. Instead, it showed an increased risk of heart disease, strokes, and breast cancer among women taking HRT.

Proper sampling methods, such as random sampling and ensuring diversity within the study population, are crucial for producing reliable and applicable results. The WHI trial's more rigorous approach to sampling provided a more accurate picture of HRT's effects. These findings highlight the dangers of sample bias, overgeneralization, and the importance of using representative samples in research.

There were some other differences between the early studies and WHI. The early studies were observational, while WHI was a randomized controlled trial that was better able to control confounding variables. I cover those concepts in Chapter 5.

Example: Colonoscopy Interval

Let's look at a more current study that might exemplify the same problem as the early HRT studies. Is this a modern equivalent?

A 2024 Swedish study has suggested that it might be safe for some individuals to wait 15 years for a repeat colonoscopy instead of the currently recommended 10 years, provided their initial results were negative. The study indicates that extending the interval between colonoscopies does not significantly increase the risk of colorectal cancer (CRC) for individuals with no initial findings. They estimate that the longer interval only causes one additional CRC death for every 1,000 people but saves 1,000 colonoscopies for other patients.

This research has sparked interest and discussion within the medical community because it could reduce the frequency of invasive procedures and associated healthcare costs. This potential reduction is significant considering the growing number of CRC cases in younger adults. If more younger people need colonoscopies, there might be an insufficient capacity to provide them for everyone. Increasing the interval length would help alleviate this potential shortfall.

While these findings are promising, significant concerns exist about the Swedish study's applicability to populations outside of Sweden, such as the United States. The Swedish healthcare system, population demographics, and lifestyle factors differ considerably from those in the U.S. These differences can impact the generalizability of the study's conclusions just like early HRT studies.

Sweden's population is more homogeneous in terms of ethnicity, primarily white, and socioeconomic status compared to the diverse population of the United States. The Swedish study participants might not accurately reflect the broader range of genetic backgrounds, health conditions, and risk factors in the U.S. population. Lifestyle and environmental factors like diet, physical activity, and exposure to risk factors like smoking and pollution can vary significantly between countries. Additionally, Sweden has a universal healthcare system providing consistent and comprehensive care to all citizens, which can lead to different health outcomes compared to the more fragmented healthcare system in the U.S.

In conclusion, the Swedish study offers valuable insights, and its findings apply to Sweden. However, authorities must exercise caution when applying them to different populations.

Is the 15-year interval appropriate outside Sweden? Given cases like the early HRT studies, I'd want researchers to conduct more studies using a more diverse sample. Applying the Swedish study's findings without accounting for these differences could lead to misguided recommendations.

Other Bias Sources

Even when researchers meticulously perform random sampling and follow all proper protocols to obtain a representative sample, various sources of sampling bias can still affect the results. Examples include survivorship bias, nonresponse bias, undercoverage bias, and attrition bias. Each can skew the results and lead to inaccurate conclusions, highlighting the importance of understanding and mitigating these pitfalls.

The following causes of sampling bias can occur even when you use probability sampling. Even if you didn't collect the data, you should assess the possibility of these bias sources affecting them.

Survivorship Bias

Survivorship bias occurs when you tend to assess successful outcomes and disregard failures. This sampling bias paints a rosier picture of reality than is warranted by skewing the average results.

It is similar to a bias you see on Facebook and other social media. On Facebook, everything looks great and everyone is doing fantastic. However, you only see a subset of photos and stories that "survive" the filtering process of what people want to post. You're not seeing all the many other less flattering pictures and narratives. Even if a researcher took a random sample of Facebook content, it's not going to represent people's lives as a whole, just their Facebook-worthy things. That situation would lead to an overly positive view of people's everyday lives.

Survivorship bias is a sneaky problem that tends to slip into analyses unnoticed. For starters, it feels natural to emphasize success, whether it's businesses, entrepreneurs, or survivors of a medical condition. We focus on and share these stories more than the failures.

But it's not that we're willfully ignoring unsuccessful cases. Even when you try to collect a full spectrum of cases, it's often easier to find the successful ones because they are usually more visible and easier to contact than unsuccessful cases. For instance, it's easier to see what users post on Facebook than to track them down and ask them about what they didn't post. Similarly, reaching and identifying long-running businesses, successful mutual funds, and living patients is simpler than the non-surviving counterparts.

Unfortunately, even when it's unintentional and occurs due to practical issues, focusing on high-performing successes and disregarding others introduces survivorship bias. After all, you're missing a significant part of the picture by not assessing the failures. Incomplete data can distort the results and lead you astray. In a nutshell, this bias is all about the fact that some subsets of the data are more difficult, or even impossible, to collect.

Survivorship bias also triggers our tendency to confuse correlation with causation. You see successful examples with particular attributes (correlation) and incorrectly assume that those attributes cause the success. However, you do not see the other cases with similar characteristics that didn't perform well.

In short, survivorship bias can produce nonrepresentative samples even when you use random sampling, causing you to jump to incorrect conclusions.

Let's dig into some examples of survivorship bias, see how it works, and learn how to avoid it. For each of these examples of survivorship bias, notice how crucial information is downplayed or even excluded entirely.

World War Two Planes

A famous and early example of survivorship bias involves planes returning from missions during World War Two. The military wanted to put armor on the aircraft to protect vulnerable spots. However, they couldn't place armor everywhere because it would be too heavy.

They looked at the bullet holes on the planes that returned, the survivors in this example. The military's first inclination was to reinforce locations with the most hits. That seems to make sense. However, Abraham Wald, a mathematician, realized that survivorship bias was at work here.

The surviving planes were hit in the observed locations and still returned. Consequently, strengthening these locations is a low priority. Instead, inferring the missing data about where the non-returning planes were hit is critical. Wald realized they needed to reinforce the locations on returning planes that were not hit. Clearly, the aircraft that got hit in those areas did not return.

Businesses and Mutual Funds

Business and financial analysts frequently assess the financial health of firms and investment return information for mutual funds. In both cases, survivorship bias plays a role. Successful, long-running businesses and mutual funds are more prominent and easier to contact than their defunct counterparts. Psychologically, we want to see how successful cases have done well.

By assessing only surviving businesses and mutual funds, analysts will record positively biased financial and investment information. They're missing an entire segment of the population—defunct companies and mutual funds. They also won't know if other firms and funds with similar characteristics as the successful ones also failed.

Given that less successful businesses are more likely to go out of business and brokerages are more apt to close less successful mutual funds, studying existing businesses and funds will likely paint a rosier picture than warranted. Be sure to find and evaluate those long-gone businesses and dropped mutual funds to get the whole picture.

Famous College Dropout Entrepreneurs

Think about famous college dropouts, such as Mark Zuckerberg, Steve Jobs, and Bill Gates, who became highly successful. These successful examples might make you think a college degree isn't beneficial. However, that's survivorship bias at work!

These famous individuals are at the forefront of media reports. You hear more about them *because* they are unusual. You're not considering the millions of other college dropouts that aren't rich and famous. You need to assess their outcomes as well.

Products and Buildings

Survivorship bias affects our perception of products and buildings that have lasted a long time. For example, you might look at an old building, car, or other item and think, "They don't make them like they used to."

However, you're basing that on relatively few survivors! To get an accurate picture, you need to assess the items that didn't survive until the present. Obviously, that's much harder. The lower-quality goods and buildings are the missing data that are long gone, skewing our picture of the overall quality of older items.

Severe Disease in Medical Studies

Survivorship bias often occurs in medical studies involving severe diseases. Younger, healthier, and more fit patients tend to survive a disease's initial diagnosis more frequently. Hence, they are more likely to join medical studies. Conversely, older, weaker patients are less likely to survive long enough to participate in studies.

Consequently, these studies overestimate successful disease outcomes because they are less likely to include those who die shortly after diagnosis.

Medical survivorship bias can occur on a more informal basis outside of studies. You'll occasionally hear an older person say they smoked all their life and are still alive. Perhaps it's not so dangerous.

You know the drill now. What are the missing data?

The smokers who die from it are not around to talk about it! That also applies to people who in accidents don't wear seatbelts or helmets.

Peer Reviewed Journals

Survivorship bias even plays a role in peer-reviewed journals. These journals tend to publish only studies with statistically significant results. However, using the standard significance level of 0.05, 5% of all studies that don't have a real effect will have statistically significant results (i.e., false positive results) by chance. Journals are much more likely to publish articles about studies with statistically significant results.

Hence, these "false positive" studies get published. When there are enough studies on a topic, there are bound to be statistically significant results, leading to articles, and that causes researchers to believe an effect exists when it doesn't.

What's missing here? There is a larger number of non-significant studies on the topic.

Journal readers will think there's an actual effect for the research question because they read several statistically significant studies about it. Unfortunately, they don't know about all the non-significant studies, which are equivalent to the non-returning planes!

Researchers need the results from studies that didn't survive the journal review process to see the whole picture. To help offset this bias, there are databases and journals dedicated to storing studies that are not statistically significant. Unfortunately, these efforts have had a relatively small impact so far. While progress is slow, these efforts are gradually gaining more support and recognition.

How to Avoid Survivorship Bias

Survivorship bias occurs because there's a selection process that makes it harder to collect data from the less successful members of a population.

These are things like the following:

- Planes that didn't survive their mission.
- Businesses and mutual funds that fail.
- College dropouts that you don't hear about.
- Medical study subjects who die before participating.
- Studies that produce non-significant results.

You consider only the population members who make it through the selection process. Samples containing only successful examples don't represent the entire population and create a distorted view.

To minimize the impact of survivorship bias, critique the process to determine if it is occurring and consider what might be missing.

If survivorship bias affects your data, find ways to draw a representative sample from the population, not just a successful subset. That process might entail more expense and effort, but you'll get better results.

Nonresponse Bias

Nonresponse bias occurs when people who do not participate in a survey or study have different characteristics or opinions than those who do. In this situation, the sample data overrepresent the subpopulations that tend to respond instead of reflecting the whole population. Even when researchers use random sampling and start with a representative sample, the response pattern can skew the data.

When respondents and nonrespondents differ, the conclusions drawn from the survey might not accurately reflect the opinions or characteristics of the whole population. This condition can distort the results.

Individuals with the following characteristics might be less likely to participate in a study:

- Less free time (e.g., having less flexible jobs or working multiple part-time jobs).
- Lower access to technology or transportation.
- Language barriers.
- Have health conditions.
- Low interest in the subject matter.
- Older.

Nonresponse bias is a common problem in survey research because getting a 100% response rate is virtually impossible. Most response rates are less than 50%, and researchers typically consider 30% "good." In other words, a survey with a reasonable response rate might still have 70% of the sample who don't respond.

Example

Researchers conducted a survey to understand opinions about recent changes to the city's public transportation system. The survey used phone calls and online forms. Afterward, the researchers found that the response rate from people above 65 is much lower than that from other age groups. They might not have responded due to difficulties using technology to access the online forms or reluctance to answer phone calls from unknown numbers.

As a result, the survey results might not accurately represent the attitudes of the entire population by underrepresenting the opinions of older people, who can have a different perspective on the changes.

Reducing Nonresponse Bias

Reducing nonresponse bias requires implementing proactive strategies that increase response rates and ensure that the sample is representative of the target population. Typically, the methods for minimizing nonresponse bias address the characteristics of non-responders I list above. You can use the following strategies while designing and conducting the survey.

Good Survey Design

A short and easy-to-understand survey design is excellent for reducing nonresponse bias because it is less likely to overwhelm or frustrate participants, making them more willing to complete the survey. As the length and complexity increase, participation tends to decrease.

Target Relevant Group

Individuals are less likely to respond when uninterested in the subject matter. Consequently, be sure you're talking to the correct people!

Track and Send Reminders

Reduce nonresponse bias by tracking nonrespondents and contacting them. Some will participate after the follow-up contacts, increasing the response rate.

Offer Incentives

Another way to minimize nonresponse bias is to use incentives to encourage participation. For example, surveys can offer participants a chance to win a prize or enter a drawing in exchange for completing the survey.

Use Multiple Contact Methods and Modes of Data Collection

One of the best ways to minimize nonresponse bias is to use multiple modes of data collection. For example, researchers can conduct the survey by telephone and online forms. This approach allows researchers to reach people who might not respond to one mode of data collection but will respond to another. You can also contact people using their native languages and employ translators.

Ensure Respondents Remain Anonymous

People value their privacy! Hence, reassuring potential participants that the study won't link their data to their identities will help reduce nonresponse bias. This assurance is crucial when dealing with sensitive topics like health conditions and social taboos.

Assessing Nonresponse Bias After Data Analysis

So, you've completed your survey or other study and suspect that nonresponse bias might be problematic. What can you do? The priority is to understand the magnitude and scope of the problem. Then, you can interpret your sample results in that context.

Start by calculating the nonresponse rate to evaluate the size of the potential problem. If participants return 30% of your surveys, you know the nonresponse rate is 100% − 30% = 70%. Unfortunately, there's plenty of potential for bias!

Then, assess late respondents because they can look like nonrespondents. Understand this group's characteristics because it gives you an idea about those who didn't respond. Are they systematically different from those who responded earlier? If they're similar in relevant attributes, you might be fine!

Additionally, use exceptional effort to follow up with some nonrespondents. Your goal is to gather at least some information from them. You just need enough data to understand how they differ from the respondents.

These steps will help you understand the direction and magnitude of your study's nonresponse bias. If you obtain enough information, you can statistically adjust your findings to factor in nonrespondents.

Undercoverage Bias

Undercoverage bias occurs when the population list from which the researchers select their sample (aka the sampling frame) does not include all population members. When that happens, the sample can-not contain the unlisted individuals, potentially producing a biased sample that only partially represents the population.

Undercoverage and nonresponse bias are similar in that both prob-lems produce samples with insufficient diversity due to a lack of par-ticipation amongst particular subpopulations. However, the causes for these two types of sampling biases differ:

- **Undercoverage bias**: The population list does not include the subpopulation. Consequently, members of that group were never in the sample.
- **Nonresponse bias**: The list includes the subpopulation and those group members were in the sample. However, they failed to participate, so the results do not include them.

In short, undercoverage bias occurs when the sampling frame does not *cover* a subpopulation. While the difference might sound like a technicality, the solutions for minimizing each type of bias differ, making it a crucial distinction.

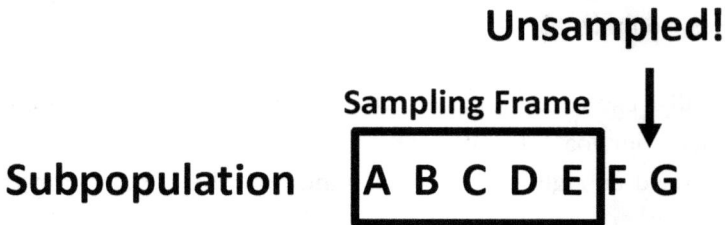

Unsampled!

Sampling Frame

Subpopulation   **A B C D E** **F G**

Undercoverage bias can have significant implications for research and decision-making. If a sample underrepresents certain groups, the re-sults might not accurately reflect the characteristics or experiences of the population. This deficiency can lead to incorrect conclusions or decisions that do not adequately address the needs of all population members. That's a problem common to all nonrepresentative samples.

Causes

Undercoverage bias occurs for two primary reasons—non-probability sampling methods and incomplete population lists.

Non-Probability Sampling Methods

Non-probability sampling methods, such as convenience sampling, tend to be biased because they do not provide an equal chance of selecting all population members for the study. Researchers who choose study participants based on proximity or ease of access cannot generalize their findings to an entire population. These methods typically do not use a complete list of the population. Many don't use a list at all!

For instance, if a researcher only collects survey responses from conference attendees, they can miss out on the perspectives of those who could not attend due to travel or financial constraints.

Incomplete Population Lists

Even probability sampling methods like simple random sampling can be susceptible to undercoverage bias if the sampling frame is incomplete. This shortcoming means that the sample might only represent part of the population because specific segments are underrepresented or not sampled. As a result, the findings might not be entirely accurate or generalizable to the larger population.

For example, suppose a researcher wants to survey the opinions of university students but only collects responses from those registered with the student government association. In that case, the sample does not accurately represent the entire population of university students. Students not involved with the student government association likely have different perspectives and opinions that the sample does not reflect.

Avoiding

To minimize the risk of undercoverage bias, you must implement strategies ensuring all segments of the population of interest have an equal chance of being selected for the study. Here are some ways to avoid undercoverage bias in your research.

Use a Comprehensive Sampling Frame

Ensure your sampling frame includes all members of the population of interest rather than just a subset. This approach can reduce the risk of undercoverage bias by ensuring that all population segments have an equal chance of being selected.

Becoming familiar with your target population is essential to capture all relevant characteristics and subgroups in your research. By understanding the nuances and diversity within your target population, you can ensure that your study is comprehensive and accurately represents the population of interest.

Consider any exclusions or limitations in your population list. These can include language barriers, geographic restrictions, or other factors that can exclude specific individuals.

Build Your Sampling Frame Using Multiple Sources

Use multiple sources to compile your population list to ensure it represents all segments. For example, if you're surveying a specific industry, you might use trade association lists, industry directories, and online databases to build your list.

Conduct Pilot Tests

Pilot tests can help you identify sampling frame and sampling method issues before launching the main study. Test runs help you address potential problems and ensure that your sample is representative of the population of interest.

By following these guidelines, you can ensure that your study's findings are accurate, reliable, and representative of the larger population.

Attrition Bias

Attrition bias in research occurs when study participants who drop out have characteristics that differ significantly from those who remain. This selective dropout can lead to skewed results and misinterpretations if the researchers don't adequately address it. This threat is higher for longitudinal studies and those with relatively high attrition rates.

Attrition bias can significantly alter your sample, resulting in a final group that markedly differs from the initial one. This shift occurs as specific segments of your original population become underrepresented in the sample. When dropouts consistently have different attributes, the remaining sample might no longer represent the original population. Consequently, the imbalanced final sample hinders your ability to generalize findings to the broader population you initially targeted.

Let's look at several examples.

Consider a long-term clinical trial testing a new drug for chronic pain management. Suppose participants with less severe pain are more likely to drop out because they perceive less need for the treatment. In that case, the remaining sample will have disproportionately more severe pain cases than the target population. Consequently, the study might underestimate the drug's effectiveness, as the data becomes skewed towards those with more severe symptoms.

Suppose a year-long study evaluates a new fitness program and begins with 500 participants of diverse ages and fitness levels. However, over time, older and less fit individuals drop out disproportionately. By the end, the remaining group skews younger and fitter, potentially over-estimating the program's effectiveness for the general population due to the lost diversity from the original sample.

In an educational experiment, if the more enthusiastic learners have more extracurricular events, they might be more likely to drop out of the study. Losing a lopsided number of devoted students can decep-tively reduce the apparent effectiveness of an educational program.

Reasons and Reducing

When delving into the reasons behind attrition bias, it's crucial to un-derstand that this form of bias doesn't occur randomly. Certain factors can systematically influence who stays and leaves a study, leading to this skewed phenomenon. Here are some key reasons why attrition bias happens:

- **Duration of Study**: Longer studies have a higher risk of par-ticipant dropout.
- **Participant Burden**: High testing frequency or intrusive methods can lead to dropout.
- **Demographic Factors**: Age, socioeconomic status, and health status can influence the likelihood of staying in a study.
- **Lack of Engagement**: A lack of perceived benefits or interest in the study can result in attrition.
- **Adverse Events**: In clinical trials, side effects or adverse events can cause participants to leave.

It's essential to employ strategic measures to mitigate the effects of attrition bias and uphold the integrity of research findings. These strategies minimize dropout rates and ensure a more balanced repre-sentation of participants throughout the study. Here are some ap-proaches for reducing attrition:

- **Effective Communication**: Regular, clear communication can keep participants engaged and informed about the study's importance.
- **Follow-Up Strategies**: Implementing reminders, follow-up calls, or emails can encourage continued participation.
- **Minimizing Burden**: Reducing testing frequency and making participation as convenient as possible can decrease dropout rates.
- **Incentives**: Providing monetary or non-monetary incentives can motivate participants to stay.
- **Keep Detailed Participant Information**: Helps ensure ongoing communication with participants, even if they relocate.

Attrition Bias After a Study

Even with preventive measures, attrition bias can still occur in research. Detecting attrition bias is a crucial step in any longitudinal study. Recognizing its presence allows researchers to address it appropriately and ensure the credibility of their findings.

To detect attrition bias, compare the characteristics of participants who drop out with those who complete the study.

For both groups, examine participants across all variables, including demographics like gender, ethnicity, age, socioeconomic status, and other relevant variables. Significant differences between groups might indicate the presence of attrition bias.

Analyze the timing and reasons for dropout. Consistent patterns, such as dropouts occurring after specific events or among certain demographic groups, can signal bias.

Employ statistical tests like logistic regression to examine whether dropout is related to treatment or outcome variables. Use dropout status as the binary dependent variable and the other variables as independent variables.

Understanding the characteristics of dropouts can help analytical methods adjust for the bias. Legitimate statistical methods to manage attrition bias exist, but they go beyond the scope of this book.

By incorporating these strategies, researchers can better understand and mitigate the effects of attrition bias, enhancing the robustness and reliability of their study's conclusions.

Symptom Based Sampling

Diagnosed conditions and referrals for treatment tend to have more severe symptoms than milder forms that are not diagnosed. This kind of sampling bias occurs in medical and psychological studies.

For example, referrals for reading comprehension problems are typically more severe. However, many more undiagnosed students might struggle with milder forms. Consequently, the sample overestimates the severity of the problem and underestimates the frequency of milder cases.

Advertising Bias

This sampling bias happens when advertising is likely to attract subjects with specific characteristics.

For example, a study that advertises a fitness improvement program is more likely to find subjects who are already motivated to get fit. Hence, the program might be more effective in this sample than in the general population.

Avoiding Sampling Bias

The previous examples of sampling bias illustrate a few of the causes. Each study has potential avenues for bias. Frequently, bias exists because particular subgroups are over or underrepresented in the sample. It's crucial to think critically about these issues.

Non-probability sampling methods are a no-brainer source of bias. Watch out for them!

In other cases, researchers might start with a representative sample but then end up with a biased sample due to lack of participation or attrition. Or, some subgroups might be underrepresented from the start because they didn't survive, are harder to contact, or were never in the sampling frame.

There are no cookie-cutter answers or guaranteed fixes. You'll need to use your subject-area knowledge and evaluate how particular subgroups might be over and underrepresented. There is a multitude of context-sensitive possibilities.

Below are some general approaches to consider:

- Use probability sampling based on a sampling frame that includes all population members.
- Identify hard-to-reach subgroups and make an extra effort to include them.
- Reduce barriers that exclude some participants, such as having flexible hours and multiple locations.
- Contact recruits who don't respond or drop out of the study.
- Find subjects using a process that doesn't entirely depend on passing a test, satisfying criteria, being diagnosed, or responding to an ad.

Research is complex and challenging. In many cases, avoiding all sources of sampling bias is impossible. However, you can take steps to minimize it. Even when you can't eliminate it, understanding sampling bias can help you better interpret the results. For example, if you advertise for an intervention, you might realize that your sample represents motivated individuals rather than the general population.

And, if you're not conducting the research yourself, these are the issues to consider when evaluating someone else's data and analysis.

Closing Thoughts

Data quality heavily depends on managing sampling methods and sources of bias. Sampling is the bedrock of data quality, and bias sources can eat away at even the strongest bedrock.

Whether you're collecting data for a new study, assessing existing data for analysis, or evaluating the analytical results of others, you must understand the context in which the data were collected. Reviewing sampling methods and how analysts addressed biases (or overlooked them) in the data collection process is critical to understanding the data's reliability.

Garbage data in will produce garbage results out. Question the data. How did the researchers collect it, and did they mitigate potential biases?

This chapter detailed various sampling methods, emphasizing how each choice impacts the data. Sampling methods are categorized into probability and nonprobability types, each with unique characteristics influencing data representation.

- **Probability Sampling**: Enhances the representativeness and generalizability of findings. However, it can be resource-intensive and challenging to implement in large populations.
- **Nonprobability Sampling**: More practical and easier to conduct but lacks the representativeness of the broader population.

With nonprobability sampling, you usually don't have a target population to which you can apply the results. Shoot, you might not fully understand the characteristics of who is participating and how they differ from the population to which you'd like to generalize.

Even when the data collection process is robust, biases like survivorship, nonresponse, undercoverage, attrition, symptom-based, and advertising bias can creep in and significantly skew results. For instance, survivorship bias might lead to overly optimistic outcomes by only considering successful cases, while nonresponse bias can produce unrepresentative data due to differential participation.

Be vigilant about various forms of sampling bias. Large sample sizes cannot correct for bias!

Can you trust the data to give you an unbiased perspective? Or do you need to be aware of potential biases?

When analysts use customer or client data stored on their server, they're engaging in a form of convenience sampling. An approach that, while practical, is fraught with potential biases.

For example, a hospital conducting a medical study using its patient data limits the sample to a specific subpopulation. This sample likely differs from the general population, introducing several types of biases. There's the risk of undercoverage bias, where the sample underrepresents particular segments of the population (e.g., healthier individuals or those with no access to healthcare). Also, hospital patients are more likely to exhibit symptoms or conditions that are not prevalent in the general population.

Such biases can skew the study's findings, leading to conclusions that might not accurately reflect broader public health trends. Big data analysts must recognize these limitations in convenience sampling and consider the potential impact on their analyses and decisions based on these data. You might have a massive database of hospital records, but the vast sample size does not eliminate or even reduce these biases.

Critically evaluate sampling methods and potential sources of bias, whether you're collecting the data yourself or evaluating existing data. Understanding and navigating these fundamental aspects of data quality is crucial for drawing sound conclusions.

There are many details, and you need to be vigilant. As I mentioned in the book's Introduction, there are many ways to get them wrong and relatively few ways to get them right. That's an ongoing theme.

CHAPTER 4

Data Quality:
Measurements

In the previous chapter, you saw how the sampling method and other aspects of acquiring and retaining participants can affect the quality of your data. This chapter remains on the Data Quality level of the Thinking Analytically pyramid. We're still focusing on data quality but moving along to issues involving *how* you measure those people or items. Not all measurement systems are created equal!

Again, garbage in, garbage out. To produce a trustworthy dataset, you must have a good sample *and* use a high-quality measurement system.

Do you trust your data?

In the intricate world of research and data analysis, measurement errors are subtle yet powerful problems that can significantly sway a study's outcomes. We'll delve into the fundamental concepts of measurement errors, laying a foundational understanding crucial for analytical thinkers.

Measurement error occurs when the measured value differs from the actual value of the quantity being measured.

There will always be a little uncertainty in our measurements. Even when you try your best, you can never measure something perfectly. It's not that we did something wrong; it's an inherent part of measuring things. In science, we call this measurement error.

Accuracy and precision are two components of data quality. By exploring these concepts, you'll understand how they shape our findings. Our exploration takes us through the contrasting terrains of systematic and random errors, shedding light on their distinct characteristics and impacts on research outcomes.

These ideas will be familiar to you because they closely relate to random error and sampling bias in the previous chapter. Are the data systematically wrong, or are the errors due to random chance? Samples and measurements can have random and systematic error.

How do researchers assess data quality? They can't just assume they have good measurements. Typically, researchers need to collect data using an instrument and evaluate the quality of the measurements. In short, they conduct an assessment before the primary research to assess data quality.

For data to be good enough to allow you to draw meaningful conclusions from a research study, they must be valid. What are the properties of good measurements?

To bring these concepts to life, I include examples from my research experiences to underscore the practical significance of understanding and addressing measurement errors. By the end of this chapter, you'll grasp the critical importance of these concepts and know how to discern and mitigate measurement errors in research.

Accuracy and Precision

Accuracy and precision are crucial properties of measurements when you're relying on data to draw conclusions. Both concepts apply to a series of measurements from a measurement system and relate to types of measurement error.

Measurement systems facilitate the quantification of characteristics for data collection. They include a collection of instruments, software, and personnel necessary to assess the property of interest. For example, a research project studying bone density will devise a measurement system to produce accurate and precise measurements of bone density.

If your project involves collecting data for analysis, your measurement system must produce data that are both accurate and precise. After all, if you can't trust the data you collect, you can't trust the results!

While people often use accuracy and precision interchangeably in everyday conversation, they have distinct definitions in statistics.

Accuracy

Accuracy assesses whether a series of measurements is correct on average. For example, if a part has an accepted length of 5mm, a series of accurate data will have an average around 5mm.

In statistical terms, accuracy is an absence of bias. In other words, measurements are not systematically too high or too low. However, accuracy tells you nothing about the distance from the target.

Please note that I've seen numerous incorrect definitions of accuracy on the Internet. Accuracy doesn't assess how close measurements are to the target. Instead, it evaluates the "correct on average" aspect. You can have data that are correct on average but fall relatively far from the proper value. That still counts as accuracy!

Accuracy relates to the central tendency of the measurements. Low accuracy corresponds with high systematic error in the measurements.

Precision

Precision indicates how close the measurements are to each other. Each measurement in a series has a component of random error. This error causes them to differ somewhat, even when measuring the same item. For example, repeatedly measuring the same 5mm part will produce a spread of values.

In this manner, precision relates to reproducibility or repeatability. How reproducible are the data when you measure the same thing multiple times? High precision measurements are closer together than low precision measurements.

However, precision tells you nothing about whether the measured values are near the correct value. Measurements can be close to each other but far from the proper value.

Examples

You might think accurate data would also be precise, and vice versa! But that's not necessarily true.

Accuracy assesses whether the measurements find the target value on average but does not indicate the distance from the target. You can have data that are correct on average but fall relatively far from the target.

For example, a project measures people's heights, but the measuring tape has too few marks. The personnel guess the values between the lines by eye and are correct on average, but there's high variation around the average. These measurements are not repeatable even though they're correct overall.

On the other hand, you can have precise measurements that are close to each other but off-target on average.

For example, imagine your bathroom scale consistently reads too high. You can take repeated weight measurements that are consistent, but they're all too high. The data are precise but inaccurate.

A *valid* measurement system is both accurate and precise. In these cases, the data are correct on average and close to the correct value. For example, if the weights from your bathroom scale center on the proper value and are close together, you have a valid scale!

Dartboard Illustrations

The classic way to represent these concepts is by using darts on a dartboard. For simplicity, I'll refer to the accepted or correct value as the target. You want your measurements to hit this target.

This dartboard represents accurate data because they average out to be on target. However, they are not precise.

The next one depicts precise data because they're close to each other. However, they are systematically off target. These data are precise but inaccurate.

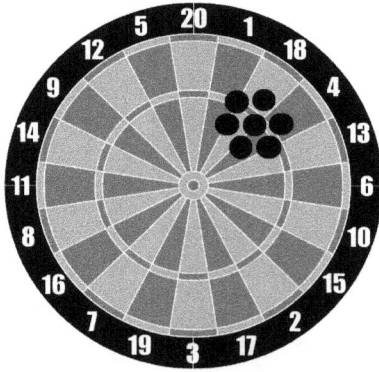

The final dartboard shows data that are both accurate and precise. The darts hit the target on average and are close together. These are the measurements you want!

Here's a handy mnemonic device for remembering which term corresponds to which concept.

- aCcuracy = Correct. Are the measurements correct on average?
- pRecision = reproducible, repeatable. Do you obtain similar values when you measure the same item multiple times?

132

Let's explore two types of measurement error and how one kind lowers accuracy and the other lowers precision.

Random Error vs. Systematic Error

Think back to Chapter 3 and how we regarded random sampling error and systematic errors. There's a similar dichotomy in measurement error:

- Random error occurs due to chance. Even if we do everything correctly for each measurement, we'll get slightly different results when measuring the same item multiple times.
- Systematic error occurs when the measurement system makes the same kind of mistake every time it measures something. Often, that happens because of a problem with a tool or how we're doing the experiment. For example, a caliper might be miscalibrated and always show larger widths than they are.

Random error lowers the precision, while systematic error lowers accuracy. Researchers must assess measurement error in scientific studies because too much of it reduces the quality of their research.

Random Error

Random error is measurement error caused by natural variability in the measurement process. It is unpredictable and occurs equally in both directions (e.g., too high and too low) relative to the correct value. Factors such as limitations in the measuring instrument, fluctuations in environmental conditions, and slight procedural variations usually cause it.

Statisticians often refer to random error as "noise" because it can interfere with the correct value (or "signal") of what you're trying to measure. If you can keep the random error low, you can collect more precise data.

For example, imagine you want to measure the height of a tree using a measuring tape. The tree's height is 10 feet, but due to variations in the measuring tape, the angle you look at it, the sun in your eyes, the wind blowing the tape, etc., you get slightly different measurements each time you measure it. The first measurement is 10.2 feet, the second is 9.9 feet, and the third is 10.1 feet. These differences are due to random error.

Unlike systematic error, we can estimate and reduce random error using statistics to analyze repeated measurements. To do this, use the same measurement device and measure the same object at least ten times. Then, find the average and the standard deviation. Although there are several ways to report the random error, a standard method is to write the mean plus or minus two times the standard deviation.

Let's return to the tree height example to illustrate random error. 10 is the correct height value for this tree.

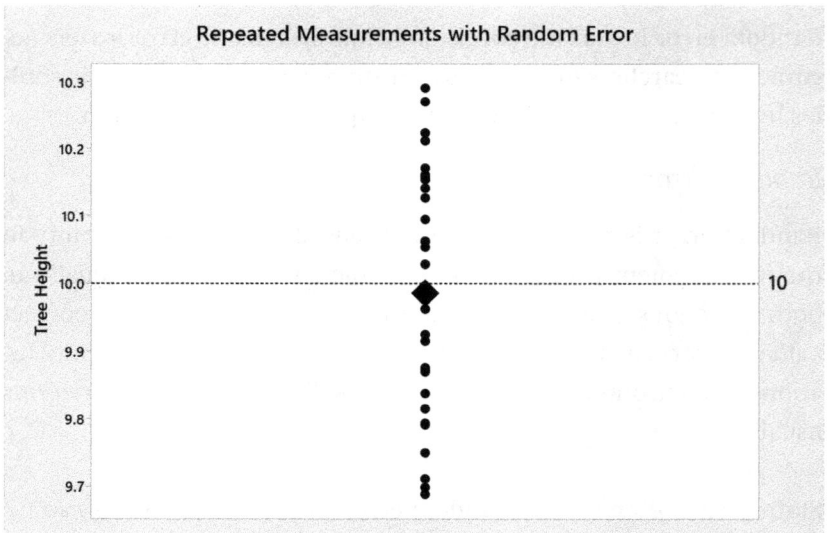

This graph shows how the measurements randomly cluster around the actual value of 10. They have no pattern. The diamond is the average

of the 30 data points, and it is close to the correct value because the positive and negative errors cancel each other out.

Random error primarily affects precision, the degree to which repeated measurements of the same thing under similar conditions produce the same result. It is unavoidable in research, even if you try to control everything perfectly. However, there are simple ways to reduce it, such as:

- **Take repeated measurements**: If you take multiple measurements of the same thing, you can average them together for a more precise result.
- **Increase your sample size**: The more data points you have, the less random error will affect your results. That's why larger sample sizes are generally better than smaller ones in terms of precision.
- **Increase the precision of measuring instruments**: Use more precise instruments or calibrate them regularly.
- **Control other variables**: Keep everything consistent so extraneous factors don't introduce random error into your measurements. By controlling all relevant variables, you minimize error sources and get more accurate results.

For example, averaging our multiple tree measurements produced a mean close to the correct value. For additional improvements, researchers can measure the tree during calm and stable meteorological conditions to reduce distracting factors. They can also use a more precise measuring tape marked with finer units. They might even use a specialized rig to hold and measure trees if they need high precision.

Systematic Error

Systematic error is a measurement error that occurs consistently in the same direction. It can be a constant difference or one that varies in relation to the actual value of the measurement. Statisticians refer to the former as an offset error and the latter as a scale factor error. In either case, a persistent factor predictably affects all measurements. Systematic errors create bias in your data.

Many factors can cause systematic error, including errors in the measurement instrument calibration, biases in the measurement process, or external factors that influence the measurement process consistently and non-randomly.

For example, imagine you want to weigh objects in an experiment. Unfortunately, the scale has a calibration error. It always shows that the weight is 1 kilogram heavier than the actual weight. Alternatively, the scale might consistently add a percentage to the correct value. Either way, this difference between the actual and measured values is systematically wrong.

That's a simple example but imagine more complex scenarios.

A survey might have a systematic error due to a cognitive bias, such as the framing effect, where the wording unduly influences the participants. The survey's language might be unintentionally prejudicial in some manner, causing people to react more negatively to survey items than they really feel.

In other cases, the expectations of the measurer and the subject can influence the measurements!

Let's return to the tree example to illustrate systematic error. The range of values in this example looks much smaller than the previous graph, but that's only due to the graph scaling.

Repeated Measurements with Systematic Error

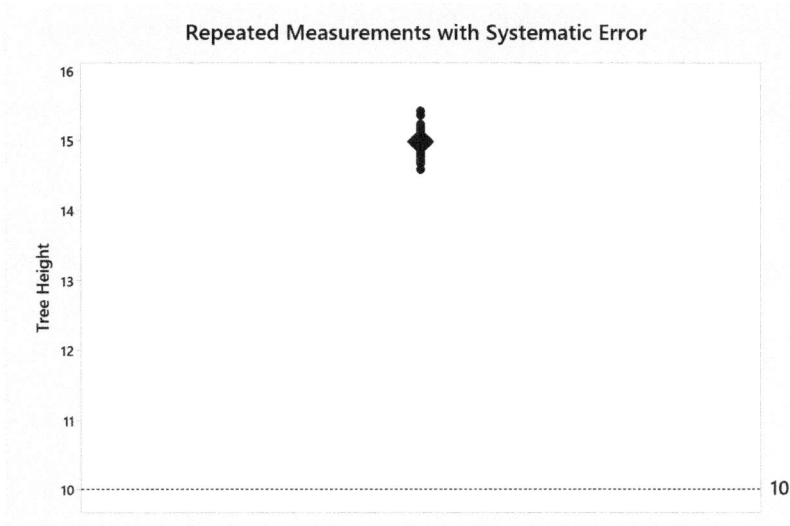

In this graph, the data points are systematically too high relative to the correct value of 10. They cluster around the wrong value. For any given measurement, you can predict that the error will be positive, making them non-random. Furthermore, the mean for these data is also wrong, unlike the random error graph. Because the errors are all positive, averaging them doesn't cancel them out.

Systematic error mainly affects accuracy by making them incorrect on average. If each measurement has a propensity to be too high, the overall average of multiple measurements will also be too high. These errors don't cancel out like random error.

To reduce systematic errors, you can use the following methods in your study:

- **Triangulation:** use multiple techniques to record observations so you're not relying on only one instrument or method.
- **Regular calibration:** frequently comparing what the instrument records with the value of a known, standard quantity

reduces the likelihood of systematic errors affecting your study.

- **Blinding**: hiding the condition assignment from participants and researchers helps reduce systematic bias caused by experimenter expectancies and cues in an experimental situation that might influence participants to behave in a certain way or provide specific responses, even if these responses do not reflect their true thoughts or behaviors.

Unfortunately, there are many possible sources of systematic error, each requiring a unique solution. So, a comprehensive list is impossible. Some instances will require extensive investigation. More on that in the next section!

Which is Worse?

Both types can be problematic, but statisticians generally consider systematic error worse than random error—the same as in the sampling context. Systematic error affects all measurements consistently in the same direction, leading to biased results. In contrast, random error affects measurements in different directions, which tends to cancel out over multiple measurements, resulting in an average value that is close to the true value. Simply increase the sample size to increase precision.

You can often identify random error by graphing the data and observing the scatter. Statistical analysis can quantify it without requiring outside information. However, systematic error is less apparent. To identify it, you need external information, such as a known correct value, as illustrated in the tree measurement example.

Notice that we only know bias exists in the tree height example because we have a standard or known value for the tree's height. Without knowing the correct value is 10, we wouldn't know all the values were too high. We'd see the scatter and assume it represents random error around the right value. Having an accepted correct value is crucial for measuring bias, as you'll see in the section about assessing accuracy and precision.

Systematic error is tricky to detect and correct. Even if you take many measurements and average them, the error remains. Unlike random error, increasing sample size or averaging does not reduce systematic error. You cannot use mathematical methods alone to eliminate systematic error or even be sure it exists.

To minimize systematic error, you can try doing these things:

- Carefully evaluate your methods and determine what might be causing the error. Then, change the procedure or conditions to fix it.
- Compare your results to studies using different equipment or methods. If their results differ from yours, it could signal systematic error in your experiment.
- Using items with known values to check your measurements. This process is called calibration.

Assessing Accuracy and Precision

The importance of assessing measurement error in any scientific or technical field cannot be overstated. It is the key to ensuring valid data upon which analysts can base sound conclusions and decisions.

Thanks to measurement error, all measurements are estimates of something's properties. Some are better than others—it's just a fact of life.

You can use measurement system analysis methods to test the accuracy and precision of your data. These analyses are specialized procedures that I'll describe in brief. Scientific experiments and quality control studies typically invest considerable time and money in assessing their measurement systems. They must trust their data before they can trust the results.

These following analyses help identify and correct potential measurement errors and ensure that data collection tools are consistently reliable over time. By understanding and applying these methods, practitioners can safeguard the integrity of their data and enhance the overall quality of their analytical results.

Gage R&R (Repeatability and Reproducibility)

Gage R&R (repeatability and reproducibility) studies test the precision of your measurement system. Specifically, it determines the sources of measurement variability using an ANOVA method. Typically, gage R&R studies tell you whether your measurements have too much variability and where to target your corrective measures. They determine how much variability originates in the devices and personnel, allowing you to identify the source of problematic variability.

This method involves multiple operators using the same measurement device to measure several items multiple times. The data collected is then analyzed statistically to separate variability into components attributable to the measurement device, operators, and actual item variation. If the procedure determines the precision is too low, it'll help you decide whether to focus improvement efforts on the instruments or people.

Suppose a factory tests the measurement system for its parts. An analyst selects ten parts representing the expected range of process variation. Three operators then measure the thickness of each part three times in a random order.

It evaluates precision by examining the consistency of measurement results when the same item is measured by the same people (repeatability) and by different people (reproducibility). Low variability in these repeated measurements indicates high precision.

Attribute Agreement Analysis

Attribute Agreement Analysis is a critical method used to evaluate the consistency and accuracy of measurement systems that rely on human judgment or subjective assessments. This analysis is vital in settings where appraisers provide subjective ratings—such as pass/fail judgments, quality ratings, or other categorizations—instead of direct physical measurements.

The ratings can be binary, nominal, or ordinal. The procedure often involves comparing appraisers' ratings to each other, to their own ratings over time, and to established standards. A reference value, or master value, is used as a known and correct rating for a standard sample, which serves as a benchmark for accuracy. Attribute Agreement Analysis is essential in maintaining reliable data collection and effective decision-making across various industries including manufacturing, healthcare, and service sectors.

These studies can assess both accuracy and precision. When the analysts include a reference value, the procedure compares the rater's assessments to the correct values to determine if they're consistently correct. It also compares the set of ratings for each item to evaluate the variability between raters.

For example, independent educational assessors grade a set of student essays for a school district's assessment using a scale of 1 to 5. The school district has determined the correct score for each essay in the sample using its grading standards. The district uses Attribute Agreement Analysis to determine whether the assessor's ratings align with its standards.

Calibration Studies

Calibration studies test the accuracy and precision of your measurement system. Typically, these studies measure items with a range of known properties multiple times and compare the measured values to known values. This process determines whether the measurements are correct on average or biased. If the data are biased high or low, you can recalibrate the device to center on the proper values.

These studies focus on accuracy. Calibration directly improves accuracy by comparing the instrument's outputs to the correct values and adjusting as needed.

A weather station calibrates its temperature sensors by comparing their readings against a mercury thermometer across various temperatures. It assesses precision by evaluating the consistency of sensor readings at the calibration points. It verifies accuracy using the average proximity of these readings to the mercury thermometer's readings at each calibration point.

Let's look at a particularly complex measurement and see how analysts go to great lengths to ensure its validity and performance. It's also a fantastic example of how some of the numbers in our lives require complex methods to produce.

Example: Measuring U.S. GDP

The GDP and other economic statistics are vital in determining the economy's health, and the policy ramifications of the quarterly estimate are the focus of much debate.

Economists define the GDP as the market value of all goods and services produced within the country during a given period. How is it possible to add the value of all goods and services for an entire quarter so quickly? Frankly, the answer is that it's impossible for such a large and complex economy, even with today's advanced technology. However, the U.S. Bureau of Economic Analysis (BEA) cranks out these measurements quarterly!

Statistical methods allow the BEA to calculate an estimate of the GDP. Let's look at how they do this to give you an idea of how much faith you want to put into the current estimate. There are other crucial economic statistics, but we'll focus on quarterly GDPs, particularly the rate of change between them.

BEA calculates these estimates to give policymakers and businesses timely and accurate estimates to devise appropriate policies. Is the economy expanding or contracting? Is the rate of change increasing or decreasing? BEA's goals are to produce good enough estimates so policymakers can rely on the early estimates as accurate indicators of the state of the economy and that policy decisions based on the early estimates don't need to be reconsidered due to revisions to GDP estimates. They can't afford to wait until everything has been counted. In short, there is a lot of pressure to produce accurate, early estimates, but little time to do it!

We generally think of account balances in concrete, bookkeeping terms. But BEA doesn't have the luxury of adding everything up. It's simply too time-consuming on such a large scale. Consequently, most of this estimation uses statistical methods and simplifying assumptions. Producing each estimate requires a broad mix of data, such as preliminary business survey results, manufacturers' shipments, various indicators, and trade industry data, among other things.

BEA calls the first quarterly estimates "Advance estimates." They release these estimates near the end of the first month after the end of the quarter. The Advance estimates are based on incomplete data, and BEA adjusts much of the data by filling in gaps or using available data as proxies for the desired measure. For example, when annual data are available but quarterly data are not, interpolation can often estimate the quarterly data. For the periods beyond those covered by annual estimates (such as the most recent quarter), extrapolation derives the quarterly estimates. Analysts often base these interpolations and extrapolations on "indicators," which are proxy data that approximate movements of the actual data. Trend projections from previous data create estimates of new data before counting even begins.

The early estimates are revised as more complete, accurate, and detailed source data become available. For example, some changes are due to the replacement of:

- Early extrapolations with newly available source data.
- Preliminary sample survey results with more complete and accurate annual and benchmark data.
- Trend projections with actual data.

Other revisions reflect changes in economic concepts and methodology necessary to provide a relevant and accurate picture of a constantly changing economy.

At the end of each of the following two months after the first estimate, BEA releases revised estimates incorporating revised and newly available monthly and quarterly data. Later, annual revisions and comprehensive revisions occur every five years.

BEA bases its accuracy standard on comparing its early and "latest" estimates, generally produced years later. (Yup, they're still trying to add everything up!) The latest revisions incorporate the most up-to-date concepts, statistical methods, and the most complete and accurate source data. A 2011 study published on BEA's Survey of Current Business concluded that the three early quarterly estimates provide an accurate picture of:

- The direction of change in real GDP 97 percent of the time
- The acceleration or deceleration of growth about 72 percent of the time
- The relative magnitude of growth more than four-fifths of the time
- The cyclical peaks before five of the six recessions from 1969–2006
- The cyclical troughs of four of the six recessions

- Trends in saving, investment, government spending, corporate profits, and other critical components of GDP

The study found no substantial revisions in key measures over time.

I wanted to examine the percent change in real GDP myself. Specifically, I wanted to compare the first Advance estimates to the latest ones. Obtaining the original quarterly Advance estimates was challenging, but thanks to Lisa Mataloni, an economist at BEA, who pointed me in the right direction! I produced the time series plot below. It compares the Advance estimates to the Latest estimates. I originally did this analysis in 2011. Hence, that's the latest data in the graph.

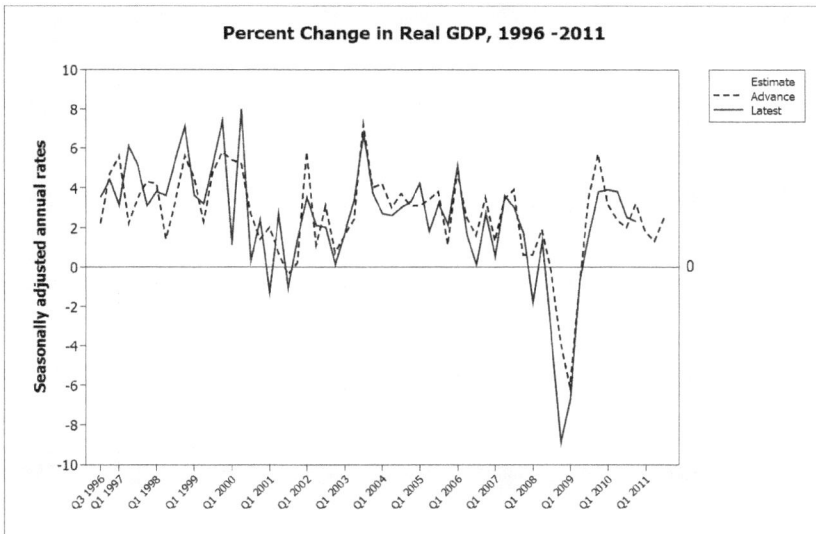

There are occasions where the Advance estimates give an incorrect impression of economic growth. When the Advance and Latest estimates are on different sides of 0, one estimate suggests growth while the other suggests contraction. This situation happened in Q1 of both 2001 and 2008. When the Advance and Latest estimates are on the same side of 0 but have slopes in opposite directions, the estimates give different indications for the change in growth or contraction. For instance, in Q1 of 1997, the Advance estimate suggests that the economy started growing faster, while the Latest estimate indicates that growth was slowing, and vice versa for the next quarter. There have also been some significant revisions. For example, in Q4 of 2008, there is the now-infamous case where the Advance estimate of -3.8% was revised sharply downward to -8.9%.

The fact that discrepancies exist is unsurprising because BEA publishes the Advance estimates just one month after the quarter ends. In contrast, the Latest estimates benefit from years of improved data, measures, and techniques. However, despite the significant constraints on the Advance estimates, the Survey of Current Business study found that the mean absolute revisions to the early quarterly change estimates for real GDP have averaged about 1% for 1983–2009. This difference is small, given the tight constraints on the Advance estimates. You can see in the time series plot that, while some points are off, the Advance and Latest estimates provide similar pictures.

Good Data is Difficult: Early On-The-Job Learning!

Now it's time for a personal story from early in my career that illustrates the difficulties of obtaining good data!

My background includes working on scientific projects as the statistician. In these positions, I was responsible for establishing valid data collection procedures, collecting usable data, and statistically analyzing and presenting the results. During the first big project in my career, I learned about the transition from learning statistics in school to applying them in a multimillion-dollar experiment and the challenges along the way.

One of the greatest things about the field of statistics is that statistical analysis provides you with a toolkit for exploring the unknown. The excitement of discovery happens after all the work to set up the experiment, collect the data, verify it, and arrange it in your statistical software. Finally, there is that moment you perform the analysis, revealing the data's meaning. That's why I love it!

Valid data is a prerequisite for using all the fantastic statistical tools. Because researchers work with questions that science has not answered, it's unsurprising that we encounter obstacles when collecting data. Often, these are novel problems. Indeed, researchers might never have collected a particular type of data before. Collecting a new data type requires scientists to learn how to collect it accurately and precisely before we can begin to answer the main research question.

On good days, you can happily think of these difficulties as creative challenges. On other days, it felt like I was just solving one problem after another. After all, we're trying to generate nice and neat data that we can analyze, but reality is messy.

My first big research project was incredibly eye-opening because I was responsible for ensuring we generated good, clean data and then analyzed it. During this early project, I became highly aware of the tremendous effort required to create good data. In contrast, data analysis classes just give you the data.

Collecting valid data is a *process*, not a single measurement event. This process requires that you have standard procedures and measurement instruments that work together in perfect harmony to produce data that you actively verify are correct. If you have a lousy process or an imperfect instrument, your data are flawed. Everything must be perfect. Being directly involved in the process's nitty-gritty details and the measurements helps ensure perfection.

This project assessed physical activity and bone density growth. I was the only full-time person on the data side. I worked with a large team of talented experts, each contributing a part of their time to this project. These experts included electrical and mechanical engineers, programmers, electricians, shop technicians, nutritionists, and a bone densitometer operator. Together, we developed and tested hardware, software, assessments, and procedures to produce an array of data types.

We also had a full-time nurse on the project who interacted with the subjects daily. She provided feedback about the suitability of using our devices and survey assessments, administered the surveys, and fitted the monitoring equipment on our subjects. Out of this milieu, I had to ensure that we produced a mountain of many different types of trust-able data. It was quite a balancing act!

Along the way, I checked and rechecked pilot data as we changed things to monitor the improvement in data quality. I also wrote stand-ard procedures to ensure consistent data collection.

It was a great learning experience. And these are the critical lessons I learned:

- Collecting good data is a process, not an event.
- You spend much time determining the best way to collect the data.
- You spend the least amount of time analyzing the data.
- You must be determined, adaptable, and willing to learn many new things.
- Don't assume anything. Check and double-check all your data streams. *Verify everything.*

You can learn data analysis in school, but there's nothing like having a multimillion-dollar project on the line to really know statistics inside and out!

The diligence required to obtain and validate data became apparent early during my time in the biomechanics lab. Imagine a young guy who's eager not to mess up. There is this nagging fear that many mistakes in research happen when you miss something or do something incorrectly at the outset, and it bites you in the derriere later. You fear you'll uncover a data problem during the analysis when it is too late to fix. I didn't want that to happen on my watch.

I quickly learned that this fear was well-founded!

Accelerometer Activity Data

As part of the bone density study, we planned to measure each subject's activity. Our subjects were to wear activity monitors for 12 hours on randomly scheduled days multiple times annually. These activity monitors use accelerometers to measure movement. We were doing this well before smartphones had this ability built in.

These devices are advanced enough to distinguish natural human actions from artificial movement types, such as riding in a car. They are also durable and easy to use — the researcher doesn't adjust anything. Crucially, other researchers validated these devices using sophisticated analyses and reported their findings in peer-reviewed journal articles.

In short, no one expected any problems with these trusted devices. I thought this would be a nice, simple place to start verifying measurements before the study to avoid problems later. It *was* an excellent place to start, but not as simple as expected!

The activity data don't translate to an exact picture of the subjects' actions. However, you can see the scores rise and fall with activity levels and compare scores to see where each subject's activity level falls within your sample of activity scores. To ensure our activity monitors worked correctly, I had pilot subjects wear the devices for a quick analysis. Sure enough, more activity produced higher readings, just as expected. So far, so good. I decided to move on to creating a standard procedure.

To collect good data, you need standard procedures for setting up and using measurement equipment. So, I wanted to establish these standards for the activity monitors. Participants wear these devices on a belt around their waist. Standardizing the device's position on the waist is a good practice—readings shouldn't be higher or lower between subjects because of inconsistent positioning.

There was no literature on the differences in positioning, so I conducted a pilot study. This time, I had the subjects wear multiple devices all around their waist. I wanted to quantify the potential risk by seeing how much the readings would vary by the device's position on each participant.

As the data came in, it first appeared that positioning was essential. The high and low readings were often different by 15%. These differences were more than I expected, but there was no other research for comparison. However, collecting more data provided insight. The positional pattern of high and low values varied from subject to subject. Finally, it became evident that while the pattern was inconsistent between subjects, it was consistent between devices.

In other words, several monitors were systematically too low by varying degrees—inaccurate data.

I contacted the manufacturer and returned the devices so they could check them out. It turned out the manufacturer had recently switched suppliers for a component, and it was causing problems.

When I re-tested the repaired monitors, the differences between positions on a subject were all less than one percent, which meant there was no practical difference between positions. All our monitors were working correctly, and position wasn't an issue. I established a consistent procedure using standard belts that fit the monitors perfectly and were infinitely adjustable to each subject's waist. I did this to prevent the monitors from flopping around due to a loose fit.

This experience was both unsettling and positive. It was unnerving because it confirmed my fears: something you miss can come back and bite you later—even when dealing with a straightforward situation. Further, some data problems are subtle. They don't show up until you check the data in several ways. It was also a positive experience because it kept me on my toes and ready for bigger challenges to come.

Load Monitor Data

This study also collected novel data using collection devices we created to measure the impact forces on a foot throughout the day. Technically speaking, we measured the ground reaction forces the foot experienced over 12 hours. It was another method of measuring activity. We theorized that bone density growth might be more sensitive to the impact forces, particularly the extra high shocks, than to the motion activity the accelerometers measured.

To collect these data, we used specialized shoe insoles that were extremely expensive and used in lab and clinical settings during short appointments to study the pressure distribution across feet during various activities. The insole manufacturer had proprietary hardware and software to calibrate the measurement system for clinical use.

However, the insoles had never been used in real-world settings and not for the 12-hour sessions we were planning. Our data was also unique because we summed the pressure to derive the total force rather than looking at the distribution of forces at different points on the foot as they did in clinical settings.

Given our new use of their insoles, we needed to devise a system. The lab designed and built a portable data collection and storage device to which we connected the insoles. We also created a calibration system and software to process the incoming data.

Because we were working with a novel measurement system, we had to be extra careful about validating our data. We did that by comparing the loads recorded by our device to a Kistler force plate, which is the gold standard for measuring ground reaction forces. We naively thought the process would be simple because clinicians had already used the insoles in clinical settings. We were just taking them out into the real world.

To make a long story short, a string of data problems delayed us from using these devices.

We could get accurate and precise data when we calibrated our system in the lab. However, when we put the insoles in the shoes and compared them to our force place, the data consistently read too low for most participants. In the trials, the degree of negative bias was strangely consistent for each person but varied between people. For example, one person would read a consistent 30% too low, another 15% too low, and another would be right on target. With the longer trials, the readings would sometimes, but not always, drift over the extended recording time.

We didn't know the reasons for any of that.

Finding the problems took a lot of time. Ultimately, we found various factors that affected the measurements, including:

- The subject's weight and foot size.
- Properties specific to the insole.
- Environmental conditions such as humidity but not temperature.
- Insole movement in the shoe.

To overcome these problems, we devised a personalized calibration routine for each subject to supplement our existing general calibration for the device. We also found a way to reduce the insoles' exposure to humidity and account for the other issues we found.

The whole process took a year longer than expected. We eventually got our data, but it was hard work.

At this point, you might be thinking, that's great you got your data, but what does it have to do with me?

The takeaway is that whether you're collecting the data, using pre-collected data for your analysis, or assessing someone else's data analysis, never take the data quality for granted. Of course, the issues I described are specific to this study. However, all data measurement projects are susceptible to real-world messiness. My story is a common one. That's why it's crucial to understand the data collection context and processes.

The further away you are from the measurements, the less effectively you can evaluate them. However, knowing that even simple data can have complexities will keep you on your toes. Do your due diligence, be critical, and ask questions!

Closing Thoughts

The past two chapters tackled the Data Quality level of the Thinking Analytically pyramid. We covered many aspects relating to how studies gather and measure their samples—collectively, that's the data collection process. Whether conducting a study or assessing one performed by others, you need to understand the entirety of the data collection. Question the data!

There are many data quality issues to consider. These include sampling methods, their pros and cons, measurement errors, random error and bias sources in both the sampling and measurement contexts, and how to mitigate them. That's a lot to get right.

Throughout this book, I develop the theme that thinking analytically successfully requires considering many details. It's easy to gloss over these details, and that's a prime culprit for making errors. This is just a forewarning that there is more to come.

I've primarily written this chapter from the perspective of someone performing these studies (the perspective I've had). However, in this era of big data, it's essential to highlight the unique problems facing people like Data Scientists and Business Intelligence Analysts. Given their roles in transforming complex data sets into actionable insights, these professionals must navigate a minefield of potential data issues. The data they analyze were often gathered for purposes other than analysis, which introduces many challenges.

In this context, data analysts must be wary of systematic errors, which various bias sources can introduce during the original data collection process and could lead to biased outcomes. Remember that while big data uses huge datasets, even a massive sample size can't correct or average out biases introduced during the sampling or measurement phases.

Analysts might see a million rows of data in their dataset, but the sheer volume will not reduce any biases that are present.

These nuances are not an academic exercise. They directly impact the analysts' ability to provide accurate, reliable insights. Their ability to identify and address these challenges is critical to the validity and utility of their findings, directly influencing decision-making processes. Being aware of potential biases and mitigating or adjusting for them is crucial.

CHAPTER 5

Experimental Design

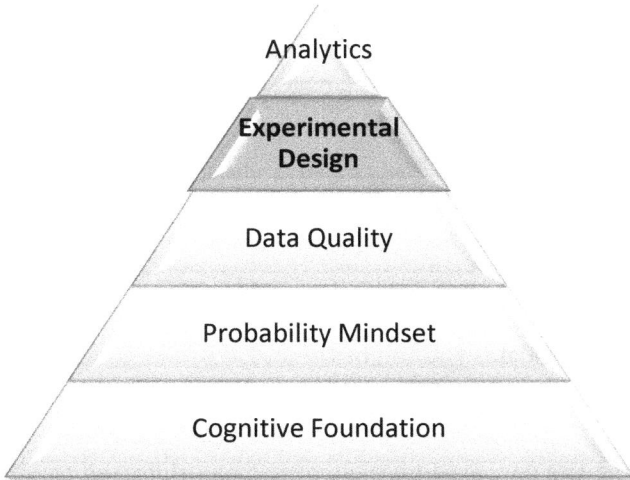

The Data Quality level of the pyramid examined the implications of how you selected and measured your sample. If you can't trust your data, you can't trust your results. Garbage in, garbage out.

This chapter takes us to the next pyramid level, Experimental Design. While your sample and data quality significantly affect the conclusions you can draw, the experimental design is the broader context in which data collection and measurements occur. This environment also affects the interpretation of the findings. It builds on the sampling and measurement processes a study has in place.

In scientific discovery, experimental design guides researchers in unraveling cause and effect. This chapter delves into experimental methodologies, contrasting their strengths and limitations. Some are better at discerning causality and avoiding biased results.

Experimentation is not without constraints. Sometimes, the ideal experiment remains just beyond reach. We'll uncover the complexities and nuances of various experimental approaches, highlighting how the quest for causality often balances on the edge of ingenuity and practicality. Understanding whether an experiment finds causal relationships, as opposed to mere correlations, is crucial. How susceptible is an experiment to confounding variables that can bias the results?

This section equips you with the necessary tools to dissect an experiment's methodology, scrutinize its control over confounding variables, and appreciate the subtleties that underpin its conclusions. By mastering these concepts, you'll be able to discern an experiment's strengths and weaknesses, empowering you to make informed decisions.

If you're in big data, you might think, I don't conduct experiments, and they don't generate my data, so why do I need this information?

Comparing the context in which your data originated to various experimental designs is crucial for understanding the types of conclusions you can make and the problems you are more likely to face. This information is directly relevant to your work.

In big data, your data were likely recorded in a process like a retrospective, observational study. Unlike structured experiments, big data often emerges not from controlled conditions but from the vast, uncharted territories of real-world interactions. This distinction underscores the constraints in drawing causal conclusions from such datasets. They also tend to use data recorded in the past for different reasons than the analysis. While big data offers a rich set of correlations, its specific context means these findings might rest on biased results and spurious correlations rather than anchored in causality.

These differences are significant because they highlight big data's limitations.

This chapter guides you through understanding experimental designs, helping you look for problems and remain vigilant. We navigate the interplay between the ideal and the attainable, shedding light on how experiments in various forms contribute to our understanding of the world.

What is Experimental Design?

An experimental design is a detailed plan for collecting and using data to identify causal relationships. Through careful planning, the design of experiments allows your data collection efforts to have a reasonable chance of detecting effects and testing hypotheses that answer your research questions.

An experiment is a data collection procedure occurring in controlled conditions to identify and understand relationships between variables. Researchers can use many potential designs. The ultimate choice depends on the research question, resources, goals, and constraints. In some fields of study, researchers refer to experimental design as the design of experiments (DOE). Both terms are synonymous.

At its most basic level, an experiment involves researchers manipulating at least one independent variable (aka factors or inputs) under controlled conditions, and they measure the changes in the dependent variable (outcomes). An effective experimental design develops a systematic procedure that increases the ability to draw meaningful conclusions from the data and reduces the interference of other variables the researchers aren't studying.

Ultimately, an experimental design helps ensure that your procedures and data will evaluate your research question effectively. Without an effective design, you might waste your efforts in a process that, for many reasons, can't answer your research question. In short, it helps you trust your results.

Experiments occur in many settings, ranging from psychology, social sciences, medicine, physics, engineering, and industrial and service sectors. Typically, experimental goals are to discover a previously unknown effect, confirm a known effect, or test a hypothesis.

Effects represent causal relationships between variables. For example, in a medical experiment, does the new medicine *cause* an improvement in health outcomes? If so, the medicine has a causal effect on the outcome.

An experimental design's focus depends on the subject area and can include the following goals:

- Understanding the relationships between variables.
- Identifying the variables that have the largest impact on the outcomes.
- Finding the input variable settings that produce an optimal result.

For example, psychologists have conducted experiments to understand how conformity affects decision-making. Sociologists have performed experiments to determine whether ethnicity affects the public reaction to staged bike thefts. These experiments map out the causal relationships between variables, and their primary goal is to understand the role of various factors.

Conversely, in a manufacturing environment, the researchers might use an experimental design to find the factors that most effectively improve their product's strength, identify the optimal manufacturing settings, and do all that while accounting for various constraints. In short, a manufacturer often aims to use experiments to improve their products cost-effectively.

In a medical experiment, the goal might be to quantify the medicine's effect and find the optimum dosage.

Developing an Experimental Design

Developing an experimental design involves planning that maximizes the potential to collect data that is both trustworthy and able to detect causal relationships. Specifically, these studies aim to see effects when they exist in the population the researchers are studying, preferentially favor causal effects, isolate each factor's actual effect from potential confounders, and produce conclusions that you can generalize to the real world.

To accomplish these goals, experimental designs carefully manage data collection, and internal and external experimental validity, which we'll cover later in this chapter. Using a well-planned experimental design gives you greater confidence that your procedures and data will produce trustworthy results.

An effective experimental design involves the following:

- Lots of preplanning.
- Developing experimental treatments.
- Determining how to assign subjects to treatment groups.

Preplanning, Defining, and Operationalizing

Due to the numerous complex objectives associated with experimental designs, there are many issues, constraints, and tradeoffs to consider before you even start collecting data. This process involves an in-depth literature review to understand the current state of scientific knowledge surrounding your research question.

This phase helps you identify critical variables, know how to measure them while ensuring accuracy and precision, and understand the relationships between them. The review can also help you find ways to reduce sources of variability, which increases your ability to detect treatment effects. Notably, the literature review allows you to learn how similar studies designed their experiments and the challenges they faced.

Operationalizing a study involves taking your research question, using the background information you gathered, and formulating an actionable plan.

This process should produce a specific and testable hypothesis using data that you can reasonably collect, given the resources available to the experiment.

Formulating Treatments

In an experimental design, treatments are variables that the researchers control. They are the primary independent variables of interest. Researchers administer the treatment to the subjects or items in the experiment and want to know whether it causes changes in the outcome.

As the name implies, a treatment can relate to medicine, such as a new medication or vaccine. But it's a general term that applies to other things, such as training programs, manufacturing settings, teaching methods, and types of fertilizers. In the bone density experiment I wrote about earlier, the treatment was a jumping exercise intervention that we hoped would increase bone density. All these treatment examples could influence a measurable outcome.

Even when you know your treatment generally, you must carefully consider the amount. How large of a dose? If you're comparing three different temperatures in a manufacturing process, how far apart are they? For my bone density study, we had to determine how frequently the exercise sessions would occur and how long each lasted.

How you define the treatments in an experiment can affect your findings and the generalizability of your results.

Assigning Subjects to Experimental Groups

A crucial decision for all experiments is how researchers assign subjects to the experimental conditions—the treatment and control groups. The control group often, but not always, is the absence of a treatment. It serves as a basis for comparison by showing outcomes for subjects who don't receive the treatment.

How your experimental design assigns subjects to the groups affects your confidence that the findings represent genuine causal effects rather than mere correlation caused by confounders. Indeed, the assignment method influences how you control confounding variables.

This issue relates to the difference between correlation and causation. Imagine a study finds that vitamin consumption correlates with better health outcomes. As a researcher, you want to be able to say that vitamin consumption causes the improvements. However, with the wrong experimental design, you can only say there is an association. A confounder, something besides the vitamins, could cause the health benefits.

Many possible experimental designs can and do fill entire books. In this overview, I'll focus on two fundamental methods for assigning participants to the experimental groups: random assignment and observational studies.

How a study assigns subjects is a crucial distinction. When you're assessing a study's results or planning to use its data, identifying its subject assignment methodology should be your first goal. That'll help you interpret the results and identify potential biases.

Random Assignment

Random assignment uses a chance process to assign subjects to experimental groups. This process requires that the experimenters can control the group assignment for all study subjects. In other words, they can control who receives the treatment and those who do not.

To illustrate how this works, let's return to the hypothetical vitamin supplement study. For our research, we must be able to assign our participants to either the control group or the supplement group. Obviously, if we can't assign subjects to the groups, we can't use random assignment.

Additionally, the random process must have an equal probability of assigning a subject to any group. For example, in our vitamin supplement study, we can use a coin toss to assign each person to either the control or supplement group. We can use a random number generator or even draw names out of a hat for more complex experimental designs.

Random assignment helps you separate causation from correlation and rule out confounding variables. As a critical component of the scientific method, experiments typically set up contrasts between a control group and one or more treatment groups. The idea is to determine whether the effect, which is the difference between a treatment group and the control group, is statistically significant. If the effect is significant, group assignment correlates with different outcomes.

However, as you have undoubtedly heard, correlation does not necessarily imply causation. In other words, the experimental groups can have different mean outcomes, but the treatment might not be causing those differences even though the differences are statistically significant.

Potential confounding variables make it difficult to state definitively that a treatment caused the difference. Confounders are alternative explanations for differences between the experimental groups. In this situation, confounding variables can be the actual cause for the outcome differences rather than the treatments themselves. As you'll see, if an experiment does not account for confounding variables, the confounders can bias the results and make them untrustworthy.

Random assignment distributes confounding properties amongst your experimental groups equally and helps eliminate systematic differences between groups when the experiment begins. This equivalence at the start increases your confidence that the treatments caused any differences you see at the end. The randomization tends to equalize confounders between the experimental groups and, thereby, cancels out their effects, leaving only the treatment effects.

Random assignment is a simple, elegant solution to a complex problem. For any given study area, there can be a long list of confounding variables that you could worry about. However, using random assignment, you don't need to know what they are, how to detect them, or measure them. Instead, use random assignment to equalize them across your experimental groups so they're not a problem.

For our vitamin study, flipping the coin tends to equalize the distribution of subjects with healthier habits between the control and treatment groups. Consequently, these two groups should start roughly equal for all confounding variables, including healthy habits! Because the groups are approximately equal when the experiment begins, if the health outcomes are different at the end of the study, the researchers can be confident that the vitamins caused those improvements.

Statisticians consider randomized experimental designs to be the best for identifying causal relationships.

Observational Studies

In some cases, randomly assigning subjects to the experimental conditions is impossible or unethical. The researchers simply can't assign participants to the experimental groups. However, they can *observe* them in their natural groupings, measure the essential variables, and look for correlations.

Observational studies let you perform research when you can't control the treatment. Unfortunately, they increase the problem of confounding variables. For these designs, correlation does not necessarily imply causation. While special procedures can help control confounders in an observational study, you're ultimately less confident that the results represent causal findings.

Consider using an observational study when random assignment is problematic. This approach allows us to proceed and draw conclusions about effects even though we can't control the condition of interest.

Examples

The following examples will help you understand when and why to use observational studies.

Researchers studying how depression affects the performance of an activity cannot assign subjects to the depression and control group randomly. However, they can have subjects with and without depression perform the activity and compare the results in an observational study.

Or imagine trying to assign subjects to cigarette smoking and non-smoking groups randomly! However, in an observational study, you can observe people in both groups and assess the differences in health outcomes.

Suppose you're studying a treatment for a disease. Ideally, you recruit a group of patients with the disease and then randomly assign them to the treatment and control group. However, it's unethical to withhold the treatment, which rules out random assignment. Instead, you can compare patients who voluntarily do not use the medicine to those who do.

In all these observational study examples, the researchers do not assign subjects to the experimental groups. Instead, they observe people already in these groups and compare the outcomes.

Types

In observational studies, researchers only observe the outcomes and do not manipulate or control factors. Despite this limitation, there are various types of observational studies.

The following designs are standard observational studies.

- **Cohort Study**: A longitudinal observational study that follows a group sharing a defining characteristic. These studies frequently determine whether exposure to risk factors affects an outcome over time.
- **Case-Control Study**: A retrospective observational study that compares two existing groups—the case group with the condition and the control group without it. Researchers compare the groups, looking for potential risk factors for the condition.
- **Cross-sectional study:** This type of study takes a snapshot of a moment in time so researchers can understand the prevalence of outcomes and correlations between variables at that instant.
- **Longitudinal Study**: A longitudinal study is an experimental design that takes repeated measurements of the same subjects over time. These studies can span years or even decades and are much more feasible as observational studies rather than true experiments.

Observational Studies vs. Randomized Experiments

Experiments excel in establishing causality, controlling variables, and minimizing the impact of confounders. However, they are more expensive, and randomly assigning subjects to the treatment groups is impossible in some settings.

Conversely, observational studies provide real-world insights, are less expensive, and do not require randomization. However, they are more susceptible to the effects of confounders, and identifying causal relationships is problematic.

Aspect	Observational Study	Random Assignment
Causality	Hard to establish	Strongly supports causality
Control of Variables	Limited or no control	High control
Real-World Insights	Strong	Limited
Cost and Time Efficiency	Cost-effective and less time-consuming	Expensive and time-intensive
Confounding Variables	Highly susceptible	Low susceptibility
Randomization	Not used	Standard practice
Longitudinal Research	Well-suited	Possible but often challenging

You probably noticed that the topics of correlation versus causation and confounding variables came up multiple times. Both issues are crucial for understanding the contrast between random assignment and observational studies. In the upcoming sections, we'll dig more deeply into both topics.

Correlation versus Causation

You've undoubtedly heard that correlation doesn't imply causation. Why is that the case, what are the differences between them, and why do they matter?

Let's first compare the definitions of correlation and causation.

Correlation means that as one variable changes, another tends to change in a specific direction. In other words, two variables move together. Positive and negative correlations exist.

- **Positive correlation**: X increases and Y tends to increase.
- **Negative correlation**: X increases and Y tends to decrease.

For example, as people's heights grow, their weight tends to increase, creating the positive correlation below.

Scatterplot of Weight kg vs Height M

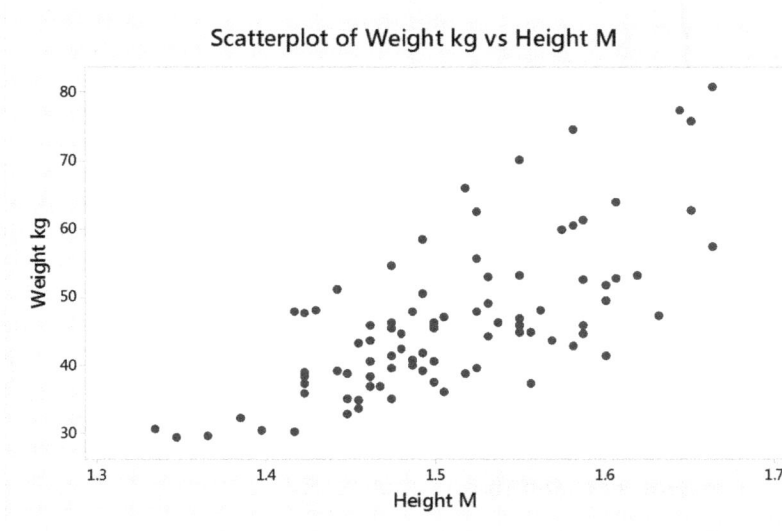

Or, as school absences increase, grades tend to decrease. That's a negative correlation.

Causation indicates that changes in one variable trigger changes in another variable. For example, increasing the dosage of a medicine *causes* the severity of the symptoms to decrease.

At first glance, those definitions certainly seem consistent, which is why they're so frequently misunderstood. What is the relationship between correlation and causation?

Correlation doesn't imply causation, but causation suggests that correlation exists. The Venn diagram shows the relationship between the two.

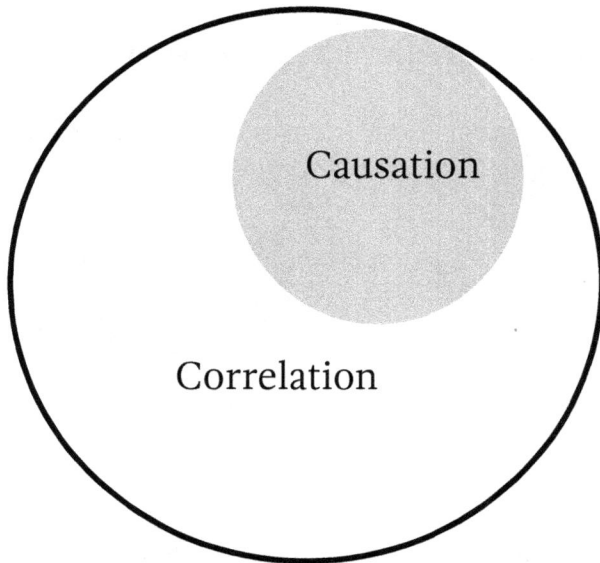

Understanding why causation implies correlation is intuitive. If increasing medicine dosage decreases the symptoms, you'll find a negative correlation between those variables. The causation creates the correlation.

Unfortunately, it's less intuitive to understand how correlation might not be causation. Let's dig into that issue.

Why Doesn't Correlation Imply Causation?

Suppose you find a positive correlation between X and Y. How could it not be causal? After all, when X goes up, Y also goes up. It sounds like cause and effect, but it might not be. Statisticians refer to a non-causal association between variables as a spurious correlation.

They exist for multiple reasons. The fact that they exist at all is why you can't be sure that correlation indicates causation. Only a subset of correlations reflects causal relationships.

Let's cover three potential explanations for non-causal correlations.

Confounders: The Third Variable Problem

A third variable can create a spurious relationship between two variables. It depends on the pattern of correlations between the two variables you're considering and a third variable.

Did you know a positive correlation exists between ice cream sales and shark attacks?

Now, ice cream sales do not *cause* an increase in shark attacks. So, what is going on?

Outside temperature positively correlates with ice cream sales and shark attack opportunities (because more people go to the beach). So, when temperature increases, both sales and attacks increase in unison, creating a spurious correlation between them. In this scenario, we call temperature a lurking variable or a confounder.

Direction of Causation

Sometimes, two variables have a causal relationship, but it's unclear which variable is the cause and which is the effect.

Researchers find a correlation between the number of hours students spend on social media and their academic grades.

Scenario 1: Spending more time on social media could distract students from their studies, leading to lower grades.

Scenario 2: Conversely, students struggling with their studies might turn to social media for escapism, meaning lower grades lead to increased social media usage.

In this situation, it's ambiguous whether social media usage causes lower grades or if the reverse is true.

Pure Chance

Random chance in sample data can produce an apparent relationship between variables. If you collect enough random samples, randomness will occasionally create the appearance of a correlation where none exists.

For these reasons, you might see a correlation in your data when there is no cause and effect.

Why Distinguishing between Them is Important

Correlation only indicates that two variables move together, but it doesn't tell us if one causes changes in the other. Relying solely on correlation can lead to misguided conclusions and ineffective or harmful actions. Establishing causality ensures that we're targeting the root cause of an issue rather than just an associated symptom.

For instance, a study by Jay and Jay (2015) in *Language Sciences* found a correlation between individuals who use taboo words (swear words) and higher levels of verbal intelligence. However, it's essential to approach such findings carefully. The correlation doesn't suggest that swearing enhances intelligence. Instead, it might indicate that individuals with a richer vocabulary, encompassing both standard and taboo words, have a more extensive linguistic repertoire to express themselves. Misunderstanding this correlation could lead to the mistaken belief that increasing one's use of swear words would boost intelligence, which is not what the study implies.

Alternatively, suppose you unknowingly find a spurious correlation between vitamins and improved health outcomes. Believing that the vitamins cause those improvements when it's merely correlation leads to poor decision-making. After all, if the vitamins don't *cause* health gains, then consuming more vitamins won't produce better outcomes despite the correlation.

We'll return to this idea later in the chapter when we examine an actual vitamin study in the scientific literature.

These examples underscore the critical importance of distinguishing between correlation and causation in decision-making.

While correlation can provide hints about potential relationships between variables, it doesn't prove that one variable causes another to change. That's an entirely different matter. They might not be causally linked. Unfortunately, confounders and spurious correlations occur frequently, and there's no statistical test for detecting them.

So, one more time, how do you distinguish between correlation and causation?

Researchers use experiments with random assignment to truly establish causality.

Conversely, observational studies are suited for finding correlations quickly and inexpensively in preliminary studies but are unsuitable for establishing causality.

Confounding Variables

In the previous sections, the concept of confounding variables came up frequently. Experiments must have a plan for addressing them. Otherwise, you can't trust their results. Let's dig into this concept more deeply. Understanding confounders is crucial whether you're assessing an experiment or working with big data, which is essentially an observational study.

A confounding variable is an unaccounted factor that relates to both a study's potential cause and effect and can distort the results. Recognizing and addressing these variables in your experimental design is crucial for producing valid findings. Confounding variables are also known as confounders, omitted variables, and lurking variables.

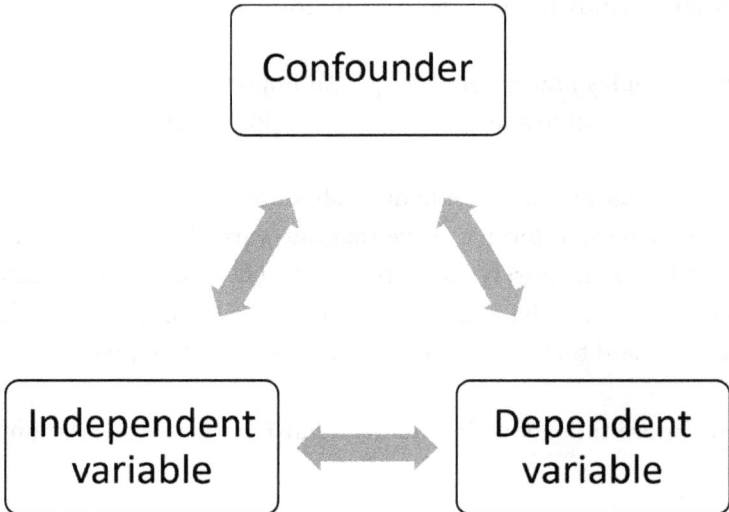

A confounding variable systematically influences both an independent (factor or input) and dependent (outcome) variable in a manner that changes the apparent relationship between them. Failing to account for a confounding variable can bias your results, leading to erroneous interpretations. This bias can produce the following problems:

- Overestimate the strength of an effect.
- Underestimate the strength of an effect.
- Change the direction of an effect.
- Mask an effect that exists.
- Create spurious correlations.

If you don't control for confounders and observe changes in your outcome of interest, the confounders might be causing the changes rather than the treatment or intervention. You don't want any of these problems!

Confounding variables bias the results when researchers don't account for them. How can variables you don't measure affect the results for variables that you record? At first glance, this problem might not make sense.

Confounding variables correlate with both the treatment and the outcome, distorting the observed relationship between them. The following two conditions must exist to create a confounding variable:

- It must correlate with the outcome.
- It must correlate with at least one explanatory or treatment variable in the experiment.

The following diagram illustrates these two conditions. There must be non-zero correlations (r) on all three sides of the triangle. X1 is the independent variable of interest, while Y is the dependent variable. X2 is the confounding variable.

Independent Dependent

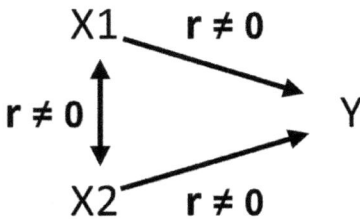

The correlation structure can cause confounding variables to bias the results that appear in your statistical output. In short, the amount of bias depends on the strength of these correlations. Strong correlations produce greater bias. If the relationships are weak, the bias might not be severe. If any correlations are zero, the extraneous variable won't produce bias, even if the researchers don't control for it.

Examples

Confounders have affected countless studies. Below are several examples of studies that have overestimated or underestimated an effect.

Exercise and Weight Loss

In a study examining the relationship between regular exercise and weight loss, diet is a confounding variable. People who exercise are likely to have other healthy habits that affect weight loss, such as diet. Without controlling for dietary habits, it's unclear whether weight loss is due to exercise, changes in diet, or both.

Power Line Proximity and Cancer Rates

Nancy Wertheimer and Ed Leeper published a report in 1979 about a relationship between childhood cancer and the "electrical current configuration" of homes in Denver, Colorado. Initial analysis attributed higher cancer rates to power line proximity.

However, later studies found that income is a confounding variable. Homes closer to power lines were predominantly in lower-income areas, where residents typically have higher cancer rates. Thus, income, rather than electromagnetic exposure from power lines, likely explains the increased cancer incidence because there is an established link between income and cancer rates.

Wine Consumption and Heart Health

Earlier studies suggested that wine consumption increases heart health. However, diet quality is a confounding variable in this subject area.

In a study conducted by Tjønneland et al. (1999) in Denmark with over 48,000 participants, researchers explored how different alcoholic beverages relate to diet quality. They found that people who drank wine tended to have healthier diets compared to those who consumed other types of alcohol. Specifically, wine drinkers were more likely to consume more fruits, fish, cooked vegetables, and salads. They also preferred using olive oil for cooking and avoided using butter or margarine on bread.

This connection suggests that wine drinkers' healthier eating habits might confound wine's observed health benefits in reducing heart disease. This finding is essential for understanding previous studies that link wine consumption to heart health.

Exercise and Bone Density: In Depth

For the bone density study I've mentioned, we measured various characteristics, including the subjects' activity levels, their weights, and bone densities, among many others. Bone growth theories suggest that a positive correlation between activity level and bone density likely exists. Higher activity should produce greater bone density.

Early in the study, I wanted to validate our initial data quickly by using simple regression analysis to assess the relationship between activity and bone density. There should be a positive relationship. To my great surprise, there was no relationship at all.

Long story short, a confounding variable hid a significant positive correlation between activity and bone density. The offending variable was the subjects' weights because it correlates with both the explanatory (activity) and outcome variable (bone density), thus biasing the results.

After including weight in the regression model, the results indicated that both activity and weight are statistically significant and positively correlate with bone density. Accounting for the confounding variable revealed the actual relationship!

The diagram below shows the signs of the correlations between the variables. Let's see how the confounder (Weight) hid the genuine relationship.

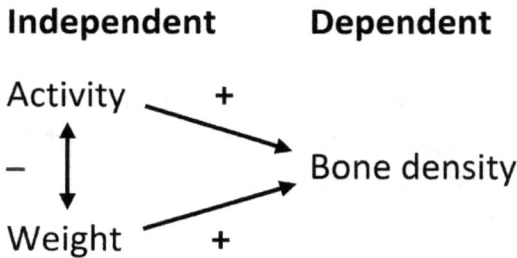

Independent   Dependent

Activity ___ +

–

Weight ___ +

Bone density

The diagram indicates that the conditions exist to allow the confounding variable (Weight) to bias the results—all three sides of the triangle have non-zero correlations. This correlation structure produces two opposing effects of activity on bone density. More active subjects get a bone density boost directly. However, they also tend to weigh less, which reduces bone density.

When I fit a regression model with only activity, the model had to attribute both opposing effects to activity alone, resulting in a zero correlation. However, when I fit the model with both activity and weight, it could assign the opposing effects to each variable separately.

Now, imagine if we didn't have the weight data. We wouldn't have discovered the positive correlation between activity and bone density. This example shows the importance of controlling confounding variables, which leads to the next section!

Reducing the Effect of Confounders

As you saw above, accounting for the influence of confounding variables is essential to ensuring the validity of your findings. Here are four methods to reduce their effects.

Restriction

Restriction involves limiting the study population to a specific group or criteria to eliminate confounding variables.

For example, in a study on the effects of caffeine on heart rate, researchers might restrict participants to non-smokers. This restriction eliminates smoking as a confounder that can influence heart rate.

Matching

This process involves pairing subjects by matching characteristics pertinent to the study. Then, researchers randomly assign one individual from each pair to the control group and the other to the experimental group. This randomness helps eliminate bias, ensuring a balanced and fair comparison between groups. This process controls confounding variables by equalizing them between groups. The goal is to create groups that are as similar as possible except for the experimental treatment.

For example, in a study examining the impact of a new education method on student performance, researchers match students on age, socioeconomic status, and baseline academic performance to control these potential confounders.

Statistical Control

Statistical control involves using analytical techniques to adjust for the effect of confounding variables in the analysis phase. Researchers can use methods like regression analysis to control potential confounders. This process requires that you measure the confounders and include them in your model.

For example, I showed how I controlled for weight as a confounding variable in the bone density study by including it in the regression model, revealing the genuine relationship between activity and bone density.

Random Assignment

Randomly assigning subjects to the control and treatment groups helps ensure that the groups are statistically similar, minimizing the influence of confounding variables.

For example, in clinical trials for a new medication, participants are randomly assigned to either the treatment or control group. This random assignment helps evenly distribute variables such as age, gender, and health status across both groups.

Of these methods, random assignment is the gold standard. Why?

For non-random assignment methods, you must be aware of all potential confounders and then be able to use restriction, matching, or statistical control to account for them. If a confounder exists that you don't know about or can't measure, it will still affect your results. What you don't know or measure can hurt you.

Conversely, randomly assigning subjects to the experimental groups works on all confounders, whether you know about and measure them or not! What you don't know or measure is much less likely to affect your results.

By incorporating these strategies into research design and analysis, researchers can significantly reduce the impact of confounding variables, leading to more accurate results.

If you aren't careful, the hidden hazards of a confounding variable can completely flip the results of your experiment!

Let's examine a specific type of randomized experiment and an observational study to see their strengths and weaknesses in action.

Randomized Controlled Trials (RCTs)

A randomized controlled trial (RCT) is a prospective experimental design that randomly assigns participants to an experimental or control group. While randomization springs to mind when discussing RCTs, other equally vital components shape these robust experimental designs. Let's look more closely at RCTs and the standard methods they employ to strengthen their results beyond randomization.

Scientists use randomized controlled trials most frequently in fields like medicine, psychology, and social sciences to rigorously test interventions and treatments.

Example

Imagine testing a new drug against a placebo using a randomized controlled trial. We take a representative sample of 100 patients. 50 get the drug; 50 get the placebo. Who gets what? It's random! We could flip a coin. For more complex designs, we could use computers for random assignment.

After a month, we measure health outcomes. Did the drug help more than the placebo?

Common Elements

Most well-designed randomized controlled trials contain the following elements.

- **Control Group**: Almost every RCT features a control group. This group might receive a placebo, no intervention, or standard care. You can estimate the treatment's effect size by comparing the outcome in a treatment group to the control group.
- **Blinding**: Blinding hides group assignments from researchers and participants to prevent group assignment knowledge from influencing results. More on this shortly!
- **Pre-defined Inclusion and Exclusion Criteria**: These criteria set the boundaries for who can participate based on specifics like age or health conditions.
- **Baseline Assessment**: An initial assessment records participants' starting conditions before the experiment begins.
- **Outcome Measures**: Clear, pre-defined outcomes, such as symptom reduction or survival rates, drive the study's goals.
- **Controlled, Standardized Environments**: Ensuring variables are measured and treatments administered consistently minimizes external factors that could affect results.
- **Monitoring and Data Collection**: Regular checks guarantee participant safety and uniform data gathering.
- **Ethical Oversight**: Ensures participants' rights and well-being are prioritized.
- **Informed Consent**: Participants must know the drill and agree to participate before joining.

- **Statistical Plan**: Detailing how statisticians will analyze the data before the RCT begins helps keep the evaluation objective and prevents p-hacking.
- **Protocol Adherence**: Consistency is critical. Following the plan ensures reliable results.
- **Analysis and Reporting**: Once done, researchers share the results—good, bad, or neutral. Transparency builds trust.

These components ensure that randomized controlled trials are rigorous and ethically sound, leading to trustworthy results.

Common Variations

Randomized controlled trial designs aren't one-size-fits-all. Depending on the research question and context, researchers can apply various configurations.

Let's explore the most common RCT designs:

- **Parallel Group**: Researchers randomly assign participants to an intervention or control group.
- **Crossover**: Participants randomly receive both intervention and control at different times.
- **Factorial**: Tests multiple interventions at once. Useful for combination therapies.
- **Cluster**: Groups, not individuals, are randomized. For instance, researchers can randomly assign schools or towns to the experimental groups.

Blinding

Blinding is a standard protection in randomized controlled trials. The term refers to procedures that hide group assignments from those involved. While randomization ensures initial group balance, it doesn't prevent uneven treatment or assessment as the RCT progresses, which could skew results.

So, what is the best way to sidestep potential biases?

Keep as many people in the dark about group assignments as possible. In a blinded RCT, participants, and sometimes researchers, don't know who gets the intervention.

There are three types of blinding:

- **Single**: Participants don't know if they're in the intervention or control group.
- **Double**: Both participants and researchers are in the dark.
- **Triple**: Participants, researchers, and statisticians all don't know.

It guards against sneaky biases that might creep into our RCT results. Let's look at a few:

- **Confirmation Bias**: Without blinding in a randomized controlled trial, researchers might unconsciously favor results that align with their expectations. For example, they might interpret ambiguous data as positive effects of a new drug if they're hopeful about its efficacy.
- **Placebo Effect**: Participants who know they're getting the 'real deal' might report improved outcomes simply because they believe in the treatment's power.
- **Observer Bias**: If a researcher knows which participant is in which group, they might inadvertently influence outcomes. Imagine a physiotherapist unknowingly encouraging a participant more because they know they're receiving the new treatment.

Blinding helps keep these biases at bay, making our results more reliable.

As you can see, randomization, blinding, and controlling for other types of biases help strengthen RCT results. A rigorous data collection and analysis plan is also in place from the start, increasing consistency and decreasing the potential for p-hacking methods that artificially produce significant results, which I'll discuss later in the book.

Observational Study: In-Depth Vitamin Example

Now, let's examine a specific observational study for comparison. Pay particular attention to its susceptibility to confounding variables.

Because observational studies don't use random assignment, confounders can be distributed disproportionately between conditions. Consequently, experimenters need to know which variables are confounders, measure them, and then use a method to account for them. It involves more work, and the additional measurements can increase the costs. And there's always a chance that researchers will fail to identify a confounder, not account for it, and produce biased results. However, if randomization isn't an option, consider an observational study.

Murso et al. (2011) used a 22-year longitudinal observational study to assess differences in death rates for subjects who used vitamin supplements regularly compared to those who did not. This study used surveys to record the characteristics of approximately 40,000 participants. The surveys asked questions about potential confounding variables such as demographic information, food intake, health details, physical activity, and, of course, supplement intake.

Because this is an observational study, the subjects decided whether to take vitamin supplements. It's safe to assume that supplement users and non-users differ in other ways. From their article, the researchers found the following pre-existing differences between the two groups:

"Supplement users had a lower prevalence of diabetes mellitus, high blood pressure, and smoking status; a lower BMI and waist to hip ratio, and were less likely to live on a farm. Supplement users had a higher educational level, were more physically active, and were more likely to use estrogen replacement therapy. Also, supplement users were more likely to have a lower energy intake, total fat, monounsaturated fatty acids, saturated fatty acids, and a higher intake of protein, carbohydrates, polyunsaturated fatty acids, alcohol, whole grain products, fruits, and vegetables."

Whew! That's a long list of differences! Supplement users were different from non-users in a multitude of ways that are likely to affect their risk of dying. The researchers must account for these confounding variables when they compare supplement users to non-users. If they do not, their results might be biased.

This example illustrates a fundamental difference between an observational study and an experiment. In a randomized experiment, the randomization would have equalized the characteristics of the subjects in the treatment and control groups. Instead, the study works with self-sorted groups that have numerous pre-existing differences.

To account for these initial differences in the vitamin supplement observational study, the researchers use regression analysis and include the confounding variables in the model. It's instructive to compare the raw results to the final, adjusted results.

Raw results

The raw differences in death risks for consumers of folic acid, vitamin B6, magnesium, zinc, copper, and multivitamins are NOT statistically significant. However, the raw results show a significant reduction in the death risk for users of B complex, C, calcium, D, and E.

However, those are the raw results of the observational study, and they do not control for the long list of differences between the groups at the beginning of the study. The results change dramatically after the regression model statistically controls for the confounding variables.

Adjusted results

Of the 15 supplements that the study tracked in the observational study, researchers found that consuming seven of these supplements were linked to a statistically significant INCREASE in death risk (p-value < 0.05): multivitamins (increase in death risk 2.4%), vitamin B6 (4.1%), iron (3.9%), folic acid (5.9%), zinc (3.0%), magnesium (3.6%), and copper (18.0%). Only calcium was associated with a statistically significant reduction in death risk of 3.8%.

In short, the raw results suggest that those who consume supplements either have the same or lower death risks than non-consumers. However, these results do not account for the supplement group's many healthier habits and attributes.

In fact, these confounders seem to produce most of the apparent benefits in the raw results because, after you statistically control the effects of these confounding variables, the results worsen for those who consume vitamin supplements. The adjusted results indicate that most vitamin supplements actually increase your death risk!

This research illustrates the differences between observational studies and experiments. Namely, the pre-existing differences between the groups allow confounders to bias the raw results, making the vitamin consumption outcomes look better than they really are.

In conclusion, if you can't randomly assign subjects to the experimental groups, an observational study might be correct for you. However, be aware that you'll need to identify, measure, and account for confounding variables in your experimental design.

Prospective vs. Retrospective Studies

One of the fundamental distinctions in experimental design is between prospective and retrospective studies. Understanding this distinction is crucial whether you're planning an experiment or assessing one performed by others because it helps identify the strengths and weaknesses inherent in each approach.

Prospective Studies

A prospective study is an experimental design that looks forward in time and observes events as they happen. Participants begin the study without having a condition of interest. Researchers gather data and take measurements at regular intervals to identify the occurrence of specific outcomes along with other related data.

For example, a prospective study might follow a group of participants and observe the onset of a disease over a certain period, focusing on identifying factors that increase or decrease disease occurrences. At the start of the study, none of the subjects have the disease of interest.

Statisticians generally consider prospective studies superior to retrospective studies because they are less susceptible to bias and confounding. In a prospective study, researchers carefully choose the methodology, variables, measurement procedures, equipment, personnel, and participants to effectively answer their research question rather than relying on whatever data happen to be available in records.

Researchers can ensure they conduct measurements and interventions in controlled, consistent conditions, minimizing potential confounders and biases by accounting for them in their experimental design. By selecting participants rather than depending on available records, a prospective study reduces the risk of selection bias, which occurs when the participants don't reflect the larger population.

Retrospective Studies

A retrospective study is an experimental design that looks back in time and assesses events that have already occurred. Researchers already know the outcome for each subject when the project starts. Instead of recording data going forward as events happen, these studies use participant recollection and previously recorded data for reasons unrelated to the project. These studies typically don't follow patients into the future.

In retrospective designs, researchers collect their data using existing records, allowing them to complete their assessment more quickly and inexpensively than a prospective study that must follow subjects over time and record data under carefully controlled conditions. However, the data might have been measured inconsistently or inaccurately because they weren't explicitly designed to be part of a study.

Researchers using a retrospective design must work with data measured for non-study reasons, which were not selected and assessed with the project's requirements in mind. Additionally, some studies might ask the subjects to remember information and use other subjective assessments, introducing various biases.

The statistical analysis for a retrospective study is frequently the same as for prospective designs. The main difference is in data quality rather than how researchers analyze the data.

Statisticians consider retrospective designs inferior to prospective methods because they tend to introduce more bias and confounding. Retrospective studies are observational studies by necessity because they assess past events, and performing a randomized, controlled experiment with them is impossible. However, they can be quicker and cheaper to complete, making them a good choice for preliminary research. Findings from a retrospective study can inform a prospective experimental design.

Advantages and Disadvantages

Prospective studies involve planning and executing data collection moving forward, allowing for greater control and reduction of biases. In contrast, retrospective studies look backward, using pre-existing data, which can introduce more variability and potential biases.

Aspect	Prospective	Retrospective
Measurements	Uses best to answer question	Uses available
Conditions	Controlled, standardized	Can vary greatly
Confounders	Choose best method to control	Not incorporated into design
Time and Cost	Time consuming and expensive	Quick and cheap
Rare outcomes	Need a large sample to find	Find in a database

Internal and External Validity

Internal and external validity relate to the findings of studies and experiments. These forms of validity are how statisticians formalize the experimental characteristics I've discussed in this chapter. We'll quickly look at them to put a button on this topic.

In a nutshell:

Internal validity evaluates a study's experimental design and methods. You must have a valid experimental design to draw sound scientific conclusions.

External validity assesses the applicability or generalizability of the findings to the real world. So, your study had significant findings in a controlled environment. But will you get the same results outside of the lab?

Internal Validity

Internal validity is the degree of confidence that a causal relationship exists between the treatment and the difference in outcomes. In other words, how well did the researchers perform the study? How likely is it that your treatment caused the group differences you observe? Are the researcher's conclusions correct? Or can changes in the outcome be attributed to other causes?

Establishing interval validity involves assessing data collection procedures, the reliability and validity of the data, the experimental design, and even things such as the setting and duration of the experiment. It could involve understanding events and natural processes outside the investigation. In other words, it's the whole thing. Does the entirety of the experiment allow you to conclude that the treatment causes the differences in outcomes?

Studies that have a high degree of internal validity provide strong evidence of causality. On the other hand, studies with low internal validity provide weak evidence of causality.

Typically, highly controlled experiments improve internal validity. Experiments with the following features tend to have the highest internal validity:

- They occur in a lab setting to reduce variability from sources other than the treatment.
- Use random sampling to obtain a sample that represents the population.
- Use random assignment to create equivalent control and treatment groups at the beginning.
- Include a control group to understand treatment effects.
- Use blinding and other protocols that reduce the influence of extraneous factors, such as knowledge about the treatment and experimenter bias.

Removing these properties, such as moving from the lab to the real world, being unable to randomize, or not having a control group reduces internal validity.

Internal validity relates to causality for a single study. For the study in question, did the treatment cause changes in the outcomes? Internal validity does not address generalizability to other settings, subjects, or populations. It only assesses causality for one study. We'll get into the generalizability issues when we talk about external validity.

Threats to internal validity are types of confounding variables because they provide alternative explanations for changes in outcomes. They are threats because they make us doubt causality. The real reason for apparent treatment effects might be these potential threats.

For example, imagine a weight loss program where the researchers measure the subjects' weights at the beginning, conduct the program, and then measure weights at the end. If the intervention causes weight loss, you'd expect to see decreases between the pretest and posttest.

However, various threats exist to attributing a causal connection between the weight loss program and weight changes. The following items are threats to internal validity.

History

An outside event occurred between the pretest and posttest that affected the outcomes and can reduce internal validity. Perhaps a fitness program became popular in town, and many subjects participated. It might be the fitness program rather than the weight loss program we're studying that caused the weight loss.

Maturation

The change between pretest and posttest scores might represent a process that occurs naturally over time and, thus, raises questions about internal validity. Imagine we are studying an educational program instead of a weight loss program. If the posttest scores are higher at the end, we might be observing regular knowledge acquisition rather than the program causing the increase. If it had been a natural process, we would have seen the same change even if the subjects had not participated in the experiment.

Testing

The pretest influences outcomes by increasing awareness or sensitivity among test takers. Suppose that the mere fact of weighing the subjects makes them more weight conscious and increases their motivation to lose weight.

Instrumentation

The change between tests is an artifact of a difference between the pretest and posttest assessment instruments rather than an actual change in outcomes. This threat to internal validity can involve a change in the instrument, different test administration instructions, or researchers using different procedures to take measurements. If the scale stops working correctly at some point after the pretest and displays lower weights in the posttest, the subjects' weights appear to decrease.

Mortality

Mortality refers to an experiment's attrition rates amongst its subjects—not necessarily actual deaths! It becomes a problem when subjects with specific characteristics drop out of the study more frequently than others. If these characteristics are associated with changes in the outcome variable, the systematic loss of subjects with these characteristics can bias the posttest results.

For example, in an educational program experiment, if the more dedicated learners have more extracurricular activities, they might be more likely to drop out of the study. Losing a disproportionate number of dedicated learners can deceptively reduce the apparent effectiveness of an educational program. This threat to internal validity is higher for studies with relatively high attrition rates.

External Validity

External validity relates to the ability to generalize the experiment's results to other people, places, or times. Scientific studies generally do not want findings that apply only to the relatively few subjects who participated in the study. Instead, they want to be able to use the experimental results and apply them to a larger population.

For example, if you're assessing an educational program, you don't want to know it's effective only for a handful of people. You want to apply those results beyond just the experimental setting and the particular individuals who participated. That's generalizability—and the heart of the matter for external validity.

Unlike internal validity, external validity doesn't assess causality and ruling out confounders.

There are two broad types of external validity—population and ecological.

Population Validity

Population validity relates to how well the experimental sample represents a population. Sampling methodology addresses this issue. Using a random sampling technique to obtain a representative sample greatly helps you generalize from the sample to the population because they are similar. Population validity requires a sample that reflects the target population.

On the other hand, if the sample does not represent the population, it reduces external validity, and you might be unable to generalize from the sample to the population.

Ecological Validity

Ecological validity relates to the degree of similarity between the experimental setting and the setting to which you want to generalize. The greater the similarity of key characteristics between settings, the more confident you can be that the results will generalize to that other setting. In this context, "key characteristics" are factors that can influence the outcome variable. Generalizability requires that the experiment's methods, materials, and environment approximate the relevant real-world setting to which you want to generalize.

Threats to external validity are differences between experimental conditions and the real-world setting. Threats indicate that you might not be able to generalize the experimental results beyond the experiment. You performed your research in a particular context, at a certain time, and with specific people. As you move to different conditions, you lose the ability to generalize. The ability to generalize the results is never guaranteed. This issue is one that you really need to think about. If another researcher conducted a similar study in a different setting, would that study obtain the same results?

The following practices can help increase external validity:

Use random sampling to obtain a representative sample from your study population.

Understand how your experiment differs from the setting(s) to which you want to generalize the results. Identify the factors particularly relevant to the research question and minimize the difference between experimental conditions and the real-world setting.

Replicate your study. If you or other researchers replicate your experiment at different times, in various settings, and with different people, you can be more confident about generalizability.

Relationship Between Them

There tends to be a negative correlation between internal and external validity. Experiments that have high internal validity tend to have lower external validity. And vice versa.

Why does this happen?

To understand the reason, consider the experimental conditions that produce high degrees of internal and external validity. They're diametrically opposed!

To produce high internal validity, you need a highly controlled environment that minimizes variability in extraneous variables. By controlling the environmental conditions, implementing strict measurement methodologies, using random assignment, and using a standardized treatment, you can effectively rule out alternative explanations for outcome differences. That produces a high degree of confidence in causality, which is high internal validity.

However, that artificial lab environment differs greatly from any real-world setting! To have high external validity, you want the experimental conditions to match the real-world setting. Observational studies are much more realistic than a lab setting. You experience the full impact of real-world variability! That creates high external validity because the experimental conditions are virtually the real-world setting. However, that type of study opens the door to confounding variables and alternative explanations for differences in outcomes—in other words, lower internal validity!

So, what's the answer?

Replication! Researchers can conduct multiple experiments in different places and use different methodologies—some true experiments in a lab and other observational studies in the field. This point reiterates the importance of replicating studies because no single study is ever enough.

As you can see, planning an experiment so you can draw valid conclusions and apply them to other settings requires a thorough assessment. Failure to plan appropriately for internal and external validity can cause your experiment or study to produce untrustworthy results!

Example: Coffee and Cancer Studies

Imagine you conduct an observational study that compares patients with and without pancreatic cancer to identify risk factors, and you find that coffee consumption corresponds to a sharp 2 to 3 times increase in pancreatic cancer risk—a much more substantial risk than smoking and alcohol consumption.

This study was real, and its results caused quite a stir. It was convincing and scary. After all, the researchers were looking at real people, their coffee habits, and cancer outcomes. The study in question was published in the *New England Journal of Medicine* by Brian MacMahon, a professor at Harvard University, in 1981.

Hopefully, after reading this chapter, you know observational studies are susceptible to confounders. And you start thinking about potential differences between the groups that might account for the apparent cancer risk.

It was a type of observational study called a case-control study, in which researchers compare the characteristics of two non-random groups, one with the outcome (pancreatic cancer) and those without it (the controls), to identify risk factors.

Fortunately for us coffee drinkers, this study was flawed, and we can continue to have our daily coffee. For starters, the participants were all hospital patients. Please recall from Chapter 3 that hospital patients don't reflect the general population, limiting the generalizability of the results. However, there are more problems with the study.

In the MacMahon study, the gastrointestinal doctors who identified and treated the pancreatic cancer cases also selected the control subjects. These patients didn't have pancreatic cancer, but they were seeing the same doctors for other severe conditions. They frequently had noncancerous gastrointestinal issues and were often advised against coffee consumption. They didn't represent the general population. This combination produced a control group that was artificially low in coffee consumption and didn't have pancreatic cancer.

The selection process created a bias in coffee drinking habits in the control group relative to the cancer patients. It appeared that people without pancreatic cancer drank less coffee, while those who drank more coffee had cancer. However, that was an artificial product of the sampling method and observational design.

Other early coffee studies linked coffee consumption with heart problems and cancer. However, these studies failed to control for a crucial confounding variable: smoking! It turns out that smoking correlates with both coffee consumption and cancer outcomes, satisfying the conditions for biased results.

In these early studies, the analyses attributed some of smoking's harmful effects to coffee consumption because smoking was not in the model. Some studies even suggested that coffee caused lung cancer. I don't know about you, but I don't inhale my coffee!

Later studies that include smoking in the model now correctly attribute those effects to smoking rather than drinking coffee. Studies using better approaches and models consistently find that coffee has protective effects and no link to cancer or heart conditions.

Research can be tricky. When you stray from randomized experiments, it gets trickier. Yet, you often need to use other methods. When assessing long-term coffee consumption, it would be challenging for researchers to randomly assign people to drink coffee or abstain from it for long enough to see whether they develop pancreatic cancer or heart conditions. So, they use observational studies and have improved their modeling over time to control confounders.

Anecdotal Evidence

No book trying to teach how to think analytically would be complete without a section about the dangers of anecdotal evidence, something we're all warned to disregard. Despite the warnings, anecdotal evidence can be surprisingly compelling. Why is that the case?

We've been discussing various types of experiments. As a method that attempts to find causal effects, I suppose anecdotal evidence fits the category loosely.

Anecdotal evidence is a story told by individuals. It comes in many forms, from product testimonials to word of mouth. It's often testimony, or a short account, about the truth or effectiveness of a claim. Typically, anecdotal evidence focuses on individual results, emotion drives the stories, and non-experts present the results.

The following are examples of anecdotal evidence:

- Wow! I took this supplement and lost a lot of weight! This pill must work!
- I know someone who smoked for decades, and it never produced any significant illness. Those claims about smoking are exaggerated!
- This anti-aging cream took years off. It must be the best!

The tricky thing about anecdotal evidence is that even when an individual story is true, it can still be entirely misleading. How does that work?

The table below shows how anecdotal evidence is the opposite of the best methods.

Data Best Practices	Anecdotal Evidence
Samples are large and representative. Using proper methodologies, they are generalizable outside the sample.	Small, biased samples are not generalizable.
Scientists take precise measurements in controlled environments with calibrated equipment.	Unplanned observations are described orally or in writing.
Confounders are measured and controlled.	Confounders are ignored.
Strict requirements for identifying causal connections.	Anecdotes assume causal relationships as a matter of fact.

A quick look at the table should convince you that anecdotal evidence is untrustworthy! However, it's even worse, thanks to psychological factors that prime us for believing these stories.

Think back to the cognitive biases in Chapter 1. Remember how the intuitive appeal of detailed narratives overrides statistical reasoning. Humans are more likely to remember dramatic, extraordinary personal stories. Throw in some emotion, and you'll likely believe the story.

Furthermore, if B follows A, our brains are wired to assume that A causes B.

Finally, anecdotal evidence cherry-picks the best stories—survivorship bias. You don't hear about all of the unsuccessful cases because people are less likely to talk about them.

When Fred tells an emotional story about taking a supplement and losing weight, we'll remember it and assume the supplement caused the weight loss. Unfortunately, we don't hear from the other ten people who took the supplement and didn't lose weight. Nor do we know how a control group fared relative to a representative sample of supplement users. We also don't know what else Fred might do to lose weight.

Collectively, these factors bias conclusions from anecdotal evidence toward abnormal outcomes (Remember the Bizarreness Effect?) and unjustified causal connections. Unfortunately, our minds are wired to believe this type of evidence. We place more weight on dramatic, personal stories.

Anecdotal stories are not necessarily fictional. Instead, they don't represent typical results, account for other factors, and have no control group.

If anecdotal evidence starts to win you over, remember that the results are not typical. And think back to the cognitive errors that make you want to believe those stories.

Closing Thoughts

In this chapter, we found that not all experiments are equally valid. Some are better than others at finding causal relationships and avoiding biased results due to confounders. I want to focus on these issues for several reasons.

For starters, it illustrates the challenges associated with performing experiments. As with sampling methods and measurements, it costs more to do the best experiments—and sometimes, it's not even possible to do the best! Science is hard. Over the decades, statisticians have developed thinking in this area. How do you assess experiments? What makes a stronger versus weaker study?

Additionally, no single experiment can prove a result definitively for any research question. Critical thinkers should consider this fact more frequently. You can't point to a single study and say something is proven.

A stringent, randomized lab-based study can show idealized results, but does it work in the real world? An observational study might show something that appears to work in the real world, but we can't completely rule out other potential explanations and possible biases (i.e., confounders). You need various supporting results from multiple studies to have confidence in any conclusions.

I've also noticed that many non-researchers need help understanding why some studies are weaker than others. The public is often surprised when different studies on the same subject produce conflicting results. How can that happen?

In the past several chapters, we've gone over many potential reasons why studies get things wrong and disagree with each other. Were they looking at different populations? Different measurements? Were some studies RCTs, while others were observational? How well did the studies control confounders? Potential biases based on their methods?

As you'll see in the upcoming chapters, there are additional statistical reasons why studies can disagree.

Big data has unique problems because it tends to be similar to retrospective, observational studies using convenience sampling. A process recorded data that was easy to obtain, often without considering future analysis. No one collected the data in the context of a planned experiment. Instead, they frequently originate from existing clients or customers. While such data is readily accessible, it might not be wholly representative of a broader population, potentially leading to biased or skewed analyses.

The original data collection methods and purpose might not align with current analysis goals. Additionally, the data's reliability or consistency across different contexts and times poses another hurdle. Changes in data collection methods over time or varying data recording standards can significantly impact the analysis' dependability.

Analysts using retrospective data must be vigilant about such limitations, understanding that the data's origins and how it was collected can significantly impact the accuracy and generalizability of their insights. They must employ critical thinking to mitigate biases and ensure their analyses are as reliable and relevant as possible. Do the data accurately represent what they were supposed to measure?

In this chapter, I've mainly focused on research studies. That's because science tends to be transparent and self-correcting. Later studies discover problems in earlier studies and publish new results. You don't generally hear about similar issues in big data because they don't publicize flaws and corrections.

Here is a case that happened to me in a big data context. Long ago, I interviewed for a Data Analyst position at a major university. They wanted to use data stored on their servers to guide their efforts to increase enrollment in their online university. A part of the interview process was to analyze the data and make recommendations to improve recruitment.

I found a factor that appeared to be a large driver of student enrollment. Unsurprisingly, they were aware of the apparent association. However, after I accounted for a confounding variable, it was no longer an important factor, and the results suggested it was better to pursue other avenues to increase enrollment.

I presented that finding during the interview, and the committee looked stunned. From the discussion, it was clear they didn't understand the concept of confounders and how they can distort the results. After all, they could see the direct correlation between the apparent factor and enrollment and thought that was the one to pursue. They couldn't understand how something else was creating the results.

Unsurprisingly, I didn't get that job. I think the interviewers thought I was a bit nuts! That's a great example of how confounders can be particularly problematic in the big data context and drive your improvement efforts in the wrong direction.

Data analysts must often work with subpar data not collected in randomized experiments. While those data can still be valuable, it's crucial to understand its limitations. Comparing the data collection process to experiments helps highlight the nature of those constraints.

In the previous chapter, I said that analysts might see a million rows in their dataset, and the results could still be biased due to sampling and measurement problems.

For this chapter, I'll add that a dataset can have a million *rows*, and the result can be biased due to a mere handful of missing *columns*. Those columns would contain confounding variables that could have enabled the analysts to correct confounder bias statistically.

Even if you're not performing experiments yourself, the information in this chapter will help you evaluate experiments you hear and read about. You can ask the following essential questions:

- How did the study assign subjects to the groups?
- Were the data purposely collected for the experiment or did the study use existing data?
- Does it control confounders? How?
- Does it minimize other forms of biases using blinding and other methods?
- Can you draw causal conclusions?
- How good of an experiment is it?

CHAPTER 6

Analytics: Basics

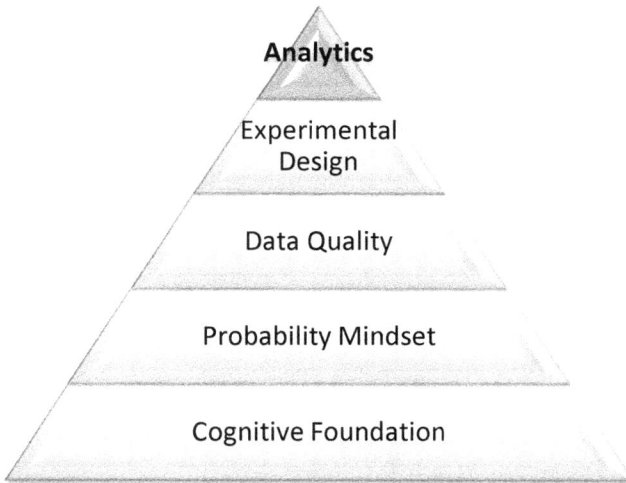

In this book, we started with our base mental capabilities for using information to make decisions and then zoomed in to see how well we assess probabilities to make decisions. Overall, we could have done better. Consequently, statisticians have developed methods and procedures for sampling populations, taking measurements, and designing experiments to overcome those natural shortcomings.

Now, we finally arrive to the data analysis. We're at the Analytics level of the Thinking Analytically pyramid! We'll be on this level for the final three chapters of the book. This level includes all types of data analytics including traditional statistics and other forms associated with Big Data. Again, this isn't a traditional statistics book. Instead of covering the types of analyses, we'll focus on how to think about analytical results, aspects to look for, and what can go wrong.

For this level, we'll generally assume that you're working with representative samples, have accurate and precise measurements, and understand any constraints of the experimental design or the observational nature of the data. In other words, there are no apparent problems so far, so what can possibly go wrong?

In statistical analysis, having good data is just the beginning. The path from having good-quality data to deriving meaningful insights is fraught with potential pitfalls—missteps that can skew your results and lead you to erroneous conclusions, even when your initial data are sound. Our focus over the final three chapters shifts to the subsequent stages of the analytical process, where even the best data and experiments can produce inaccurate results.

By understanding these pitfalls, data analysts can better navigate the intricacies of analysis, ensuring that their conclusions are not just supported by high-quality data but also robustly derived and correctly interpreted. My goal is to arm you with the knowledge to spot and avoid these common errors, enhancing the integrity and reliability of your results. You can use this information if you're assessing analyses by others or performing it yourself.

This chapter delves into the common, yet often subtle, errors that arise not from the quality of the data itself but from how analysts handle, interpret, and present it. From the fundamental challenges of dealing with small datasets, outliers, and missing data to more complex issues, such as the misinterpretation of p-values and the deceptive simplicity of graphical representation, we explore a range of scenarios where good data can lead to bad science.

These issues seem so simple, but mishandling them can substantially distort your results—and I'll show you some real cases where that happened.

Outliers

Outliers are unusual values in your dataset, and they can distort statistical analyses and violate their assumptions. It's a simple concept—they are values that are notably different from other data points, and they can cause problems in statistical procedures.

To demonstrate how much a single outlier can affect the results, let's examine the properties of an example dataset. It contains 15 height measurements of human males. One of those values is an outlier. Here's the dataset.

Height M
1.5895
1.6508
1.7131
1.7136
1.7212
1.7296
1.7343
1.7663
1.8018
1.8394
1.8869
1.9357
1.9482
2.1038
10.8135

	With Outlier	Without Outlier	Difference
Mean	2.4m (7' 10.5")	1.8m (5' 10.8")	0.6m (~2 feet)
S.D.	2.3m (7' 6")	0.14m (5.5")	2.16m (~7 feet)

From the table, it's easy to see how a single outlier can distort reality. A single value changes the mean height by 0.6m (2 feet) and the standard deviation by a whopping 2.16m (7 feet)! Hypothesis tests that use the mean with the outlier are off the mark. And, the much larger standard deviation will severely reduce statistical power.

While the concept of outliers seems straightforward, figuring out what to do with them is more complicated. Unfortunately, all analysts will confront them. Given the problems they can cause, it might seem best to remove them. But that's not always the case. Removing outliers is legitimate only for specific reasons.

Outliers can provide helpful informative about the subject area and data collection process. It's essential to understand how outliers occur and whether they might happen again as a normal part of the process or study area. Unfortunately, resisting the temptation to remove outliers inappropriately can be difficult. Outliers increase the variability in your data, which decreases statistical power. Consequently, excluding outliers can cause your results to become statistically significant.

The proper action depends on what causes the outliers. In broad strokes, there are three causes for outliers—data entry or measurement errors, sampling problems and unusual conditions, and natural variation.

Data Entry and Measurement Errors

Errors can occur during measurement and data entry, and typos can produce weird values. Imagine that we're measuring the height of adult men and gather the following dataset.

In this dataset, the value of 10.8135 is clearly an outlier. Not only does it stand out, but it's an impossible height value. Examining the numbers more closely, we conclude the zero might have been accidental. Hopefully, we can either return to the original record or remeasure the subject to determine the correct height.

These types of errors are easy cases to understand. If you determine that an outlier value is an error, correct the value when possible. That can involve fixing the typo or remeasuring the item or person. If that's not possible, you must delete the data point because you know it's an incorrect value.

Sampling Problems

Inferential statistics use samples to draw conclusions about a specific population. Studies should carefully define a population and then draw a random sample from it specifically. That's the process by which a study can learn about a population.

Unfortunately, your study might accidentally obtain an item or person not from the target population. There are several ways this can occur. For example, unusual events or characteristics can occur that deviate from the defined population. The experimenter might measure the item or subject under abnormal conditions. In other cases, you can accidentally collect an item that falls outside your target population, and thus, it might have unusual characteristics.

Let's bring this to life with several examples!

Suppose a study assesses the strength of a product. The researchers define the population as the output of the standard manufacturing process. The typical process includes standard materials, manufacturing settings, and conditions. If something unusual happens during a portion of the study, such as a power failure or a machine setting drifting off the standard value, it can affect the products. These abnormal manufacturing conditions can cause outliers by creating products with atypical strength values. Products manufactured under these unusual conditions do not reflect your target population of products from the normal process. Consequently, you can legitimately remove these data points from your dataset.

During the bone density study, I noticed an outlier in a subject's bone density growth. Her growth value was unusual. The study's subject coordinator discovered that the subject had diabetes, which affects bone health. Our study aimed to model bone density growth in pre-adolescent girls with no health conditions affecting bone growth. Consequently, we excluded her data from our analysis because she was not a member of our target population.

If you can establish that an item or person does not represent your target population, you can remove that data point. However, you must be able to attribute a specific cause or reason for why that sample item does not fit your target population.

Natural Variation

The previous outlier causes are problematic issues. They represent different types of problems that you need to correct. However, natural variation can also produce outliers—and it's not necessarily a problem.

All data distributions have a spread of values. Extreme values can occur, but they have lower probabilities. If your sample size is large enough, you will obtain unusual values. In a normal distribution, approximately 1 in 340 observations will be at least three standard deviations from the mean. However, random chance might include extreme values in smaller datasets. In other words, the process or population you're studying might produce weird values naturally. There's nothing wrong with these data points. They're unusual but are a normal part of the data distribution.

For example, I fit a regression model that uses historical U.S. Presidential approval ratings to predict how later historians would ultimately rank each President. It turns out a President's lowest approval rating predicts the historian ranks. However, one data point severely affects the model. President Truman doesn't fit the model. He had an abysmal lowest approval rating of 22%, but later historians gave him a relatively good rank of #6. If I remove that single observation, the R-squared increases by over 30 percentage points! The model appears to fit the data much better.

However, there was no justifiable reason to remove that point. While it was an oddball, it accurately reflects the potential surprises and uncertainty inherent in the political system. If I remove it, the model makes the process appear more predictable than it really is. Even though this unusual observation is influential, I left it in the model. Removing data points only because it produces a better fitting model or statistically significant results is bad practice.

If the extreme value is a legitimate observation that is a natural part of your study population, you should leave it in the dataset.

Example: Public Debt and Economic Growth

The following example illustrates how improperly handling outliers can lead to misleading conclusions. In this case, it helped drive widespread policy decisions and highlights the necessity for meticulous accuracy and transparency in data analysis. These are necessary to ensure that policy recommendations are based on reliable findings.

In their influential 2010 paper, "Growth in a Time of Debt," economists Carmen M. Reinhart and Kenneth S. Rogoff investigated the complex relationship between elevated public debt levels and economic growth.

Leveraging a vast new dataset encompassing 44 countries across 200 years, they concluded that while the link between growth and debt was relatively weak under normal conditions, public debt exceeding about 90 percent of GDP correlated with significantly lower growth rates. This finding suggested a stark drop-off in growth at high debt levels, a pattern observed across both emerging markets and advanced economies. It challenged prevailing economic theories that higher debt levels in developed nations are manageable and not detrimental to growth.

These conclusions had profound implications, influencing a broad spectrum of global austerity measures and fiscal policies. In the aftermath of the 2008 financial crisis, many policymakers and economic advisors referenced their findings to justify austerity measures. Reinhart and Rogoff's findings that high debt levels could severely hinder economic recovery and growth helped shape these discussions and policy decisions.

However, the integrity of these findings came under fire in 2013. The pivotal critique by Herndon, Ash, and Pollin exposed several critical errors in the original study, notably the mishandling of outliers and data exclusions, which significantly skewed the results. Specifically, Reinhart and Rogoff excluded data from several countries like Australia, Canada, and New Zealand during periods when these countries had high debt but also high growth. This selective exclusion of data points dramatically skewed the average growth calculations for high debt levels.

These data points showed that high debt levels are not universally associated with low growth. When Herndon et al. correct these exclusions, they found that the average real GDP growth rate for countries with public debt over 90% of GDP was actually 2.2%, not −0.1%, as claimed by Reinhart and Rogoff. This correction suggests that high debt levels do not necessarily stifle economic growth as dramatically as previously thought.

These nuts-and-bolts decisions can dramatically affect the results. The initial study's flawed findings affected policy discussions. Fortunately, science tends to be self-correcting over time.

Guidelines for Dealing with Outliers

Sometimes, as the previous example shows, it's best to keep outliers in your data. They can capture valuable information that is part of your study area. Retaining these points can be difficult, particularly when it reduces statistical significance. Removing outliers can be extremely tempting because it tends to increase statistical significance. Outliers inflate the variability, which reduces a statistical test's ability to detect an effect.

However, excluding extreme values solely due to their extremeness can distort the results by removing information about the variability inherent in the study area. You're forcing the subject area to appear less variable than it is in reality.

As an analyst, you'll run across various methods for detecting outliers, ranging from boxplots and other graphs to Z-scores. However, never remove a data point simply because a statistical method flagged it as an outlier. Use your understanding of the data's context and the study's goal to guide your decision.

When considering removing an outlier, you must evaluate whether it appropriately reflects your target population, subject area, research question, and methodology. Did anything unusual happen while measuring these observations, such as power failures, abnormal experimental conditions, or anything else out of the norm? Is there anything substantially different about an observation? Did measurement or data entry errors occur?

If the outlier in question is:

- A measurement error or data entry error, correct the error if possible. If you can't fix it, remove that observation because you know it's incorrect.
- Not a part of the population you are studying (i.e., unusual properties or conditions), you can legitimately remove the outlier.
- A natural part of the population you are studying, you should not remove it.

When you decide to remove outliers, document the excluded data points and explain your reasoning. You must be able to attribute a specific cause for removing outliers. Another approach is to perform the analysis with and without these observations and discuss the differences. Comparing results in this manner is particularly useful when you're unsure about removing an outlier and when there is substantial disagreement within a group over this question.

Proper treatment of outliers is a fundamental aspect of data analysis that requires careful consideration because it can change the nature of the results. Whether you are performing the analysis or reviewing someone else's work, it is critical to discern how outliers have been handled and to understand the implications of these decisions on the study's conclusions.

The temptation is often to remove outliers because it usually helps make the results statistically significant. However, removing outliers to enhance statistical significance creates misleading results that don't reflect reality. As we will explore in an upcoming section on p-hacking, preserving the integrity of your data analysis involves more than just achieving statistical significance—it requires a thoughtful approach to include or exclude data points based on sound reasoning and appropriate statistical techniques.

Missing Data

Missing data refers to the absence of data entries in a dataset where values are expected but not recorded. They're the blank cells in your data sheet. Missing values for specific variables or participants can occur for many reasons, including incomplete data entry, equipment failures, or lost files. When data are missing, it's a problem. However, the issues go beyond merely reducing the sample size. In some cases, they can skew your results.

Data gaps can significantly impact research integrity because they fail to represent the values intended for measurement. Understanding the root cause of these gaps is crucial because it determines whether and how to address them.

Missing data are not all created equal. Various types have distinct impacts on your dataset and the conclusions you draw from your analysis. Furthermore, the extent to which absent values affect the study results largely depends on the type. These types require different strategies to maintain the integrity of your findings.

Let's examine three types of missing data, using examples to illustrate how they might appear in real-world datasets and affect your analysis. We'll go from the best to the worst kind.

Missing Completely at Random (MCAR)

When data are missing completely at random (MCAR), the likelihood of missing values is the same across all observations. In other words, the causes for the missing data are entirely unrelated to the data itself and affect all observations equally. Consequently, you can disregard the potential for the bias that occurs with other kinds of absent values.

For the bone density study I worked on, we measured the subjects' activity levels for 12 hours with accelerometers and load monitors. Invariably, those monitors would fail randomly, and we'd lose some data. Those data are MCAR because all observations had an equal probability of containing missing values.

Fortunately, when your data gaps are MCAR, you can usually ignore them.

When data are Missing Completely at Random, their absence is independent of any measured or unmeasured variables in the study. This randomness means that the missing data are less likely to introduce bias related to the data's distribution, and you can often ignore them without distorting the analysis.

MCAR data reduces sample size and the precision of the sample estimates but tends not to introduce bias. Regular statistical hypothesis
testing will compensate for these random losses by adjusting the results to reflect the reduced sample size, thus preserving the study's
integrity.

Missing at Random (MAR)

Despite its name, MAR occurs when the absence of data is *not* random.
The probability of missing data is *not* equal for all measurements.
They're more likely for some observations than others. However,
measurements of observed variables predict the unequal probability
of missing values occurring. Crucially, those probabilities don't relate
to the missing information itself. Hence, statisticians say that the data
gaps correlate with observed values and not the unobserved (missing)
values.

For instance, consider a medical study tracking the effects of a new
medication. If patients from a particular region are less likely to complete follow-up visits — perhaps due to longer travel distances — their
follow-up data would be missing. If the dataset includes the patients'
geographical information, these missing data are MAR. The missingness depends on the observed geographic location but not directly on
the unobserved follow-up outcomes themselves. By acknowledging
the role of geography in the availability of follow-up data, researchers
can adjust their analyses to better estimate the medication's effects
across all regions.

Analysis of MAR missing data can produce biased results when analysts don't correctly handle them. This bias occurs because missing
values systematically differ from observed values, changing the properties of your sample. It is no longer a representative sample.

However, despite being non-random, MAR is a middle ground where
your results can be unbiased when you use the correct methods. If you
can use the observed variables to predict the absent values, you can

consider the missing data to be MCAR. Modern techniques for handling absent values often begin with the assumption of MAR, as it allows for more nuanced analyses that can account for observed patterns in the dataset.

Missing Not at Random (MNAR)

This type occurs when the probability of missing data relates to the absent values themselves, indicating a deeper issue within the dataset. Hence, it's a problem because you can't understand and model it.

In health surveys, individuals with more severe symptoms might be less likely to report their health status. This pattern creates a dataset where sicker individuals are underrepresented. These missing values are MNAR because only the symptom severity variable can predict its own missing values, not other variables.

Missing Not at Random (MNAR) is the most challenging type of missing data because it occurs when the absence of data directly relates to the missing values themselves. This situation can introduce significant biases because the absent values are systematically different from the ones you record. For instance, if lower-income people are less likely to report their earnings, analysis of these data will likely overestimate the average income.

You might be unable to analyze MNAR data without producing biased results. And, unlike MAR, you can't correct the bias using your observed variables. In this case, you should critically evaluate your results and compare them to other studies to assess the potential for bias and its degree.

How to Handle Missing Data

Researchers must decide on the best strategy to ensure their analysis remains robust and meaningful when dealing with missing data.

You typically have three options: accept, remove, or recreate them through imputation.

- **Accepting**: Leave the blank cells in your dataset and analyze.
- **Deletion**: Remove data points or records that have missing values. There are two primary methods:
    - **Listwise**: This technique removes an entire record when any value is missing. It's straightforward and ensures that only complete cases are analyzed.
    - **Pairwise**: Unlike listwise deletion, pairwise deletion uses all available data by analyzing pairs of variables without missing values. This method includes more data points in specific statistical analyses but likely has unequal sample sizes for different pairs.
- **Imputation**: Fills in missing data with estimated values. The simplest form replaces absent values with the variable's mean, median, or mode. More sophisticated methods, like regression imputation, predict missing values based on related information in the dataset.

Accepting missing data is best for MCAR because they are unlikely to bias your results.

The deletion methods simplify the data handling process but reduce the sample size. Critically, deletion can introduce bias when the absent values are not MCAR.

Imputation helps maintain statistical power by estimating missing values and addressing reduced sample sizes. However, it risks introducing bias if the calculated values do not accurately reflect the correct values. Choosing the proper imputation method is crucial, as incorrect assumptions can result in misleading analysis outcomes.

For MAR data, advanced techniques, such as regression or multiple imputation, can produce unbiased estimates. Consequently, they offer a significant advantage over the deletion methods.

Note: Using a measure of central tendency to replace missing values will still yield biased results for MAR data.

Navigating missing information is an essential skill in statistical analysis. By understanding the types of missing data and implementing strategies to manage them, researchers can ensure more accurate and reliable outcomes. Effective handling of absent values enriches the quality of your analysis and bolsters the credibility of your findings in the broader research community.

Remember, the goal is to handle missing data, anticipate and mitigate its occurrence, and ensure your dataset is representative and comprehensive.

Example: Space Shuttle Missing Data

On January 28, 1986, the space shuttle *Challenger* exploded during launch, largely due to missing data.

The shuttle's solid rocket boosters (SRBs) have O-rings that seal the booster's joints, preventing the burning fuel from escaping. Engineers have data relating to the O-rings from launches at various launch temperatures. They recorded the number of incidents from 21 flights. Each shuttle has six O-rings, and the engineers assessed the number showing distress.

The morning launch was an unusually cold 31F, causing NASA to assess safety on the eve of the launch. Engineers and managers convened to deliberate on the implications of launching at that temperature. At the heart of their discussion was a concern about the O-rings. Based on their material properties, there were solid engineering grounds to believe that colder conditions could raise the risk of malfunction. This consideration was crucial in guiding their decision-making process.

Their initial analysis used data from only the launches where some of the O-rings showed distress. The analysis indicated there was no relationship between launch temperature and part distress. The graph below displays the actual data. These data are publicly available to download from the University of California, Irvine, Machine Learning Repository.

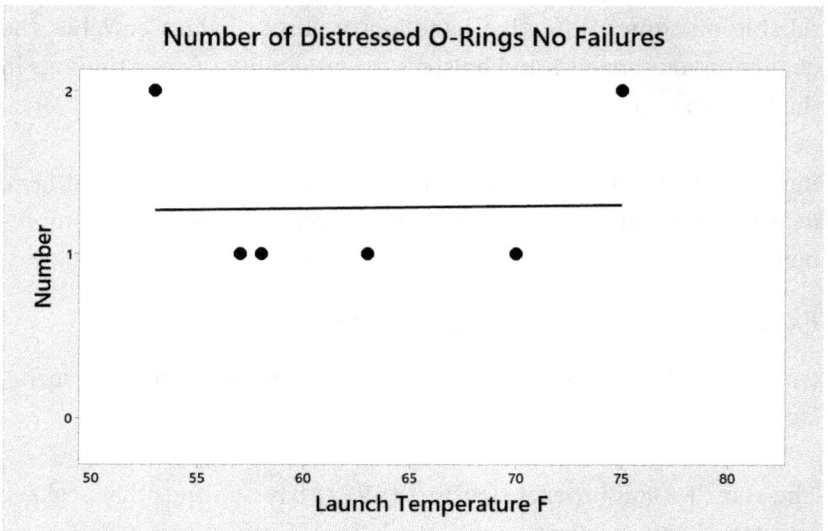

The flat trend line in the graph suggests no relationship exists between launch temperature and the number of incidents. Based on this analysis, they decided to proceed with the launch.

Unfortunately, their analysis did not include data from flights with no failures. Those data existed but were missing from their analysis. The decision to exclude those data likely cost all seven astronauts their lives.

When you include all data, launches with and without distressed parts, the relationship between launch temperature and the number of incidents becomes evident, as shown in the following graph.

Number of Distressed O-Rings All Launches

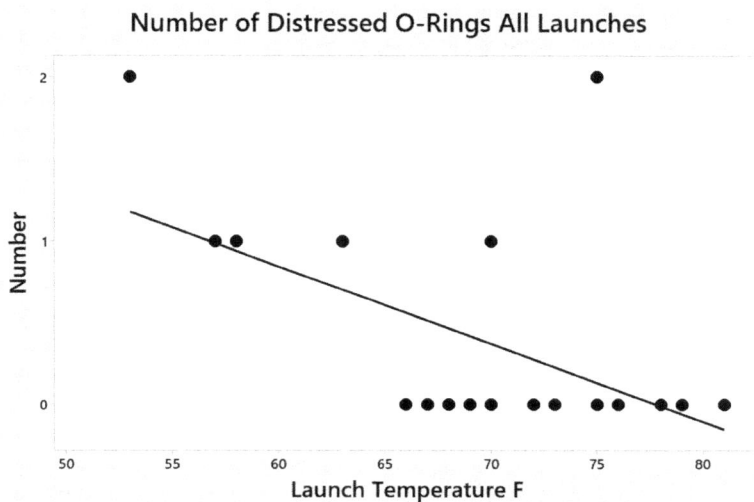

The trend line clearly shows that as temperature decreases, the number of failures increases.

Additionally, it's never good to extrapolate far outside the range of your dataset. The launch was at 31F, but the data only covered 53-75F. What if the trendline turns upward even more sharply? You'd never know because you don't have the data for those temperatures. That's another reason to call off the launch. The trendline's slope suggests more failures at temperatures below 53F, but the slope could steepen.

If we do extrapolate, a negative binomial regression model predicts that there will be five incidents at 31F. That's more than twice any other shuttle flight and nearly all six O-rings.

This example shows how easily smart people can make a mistake. Excluding the data at the time might have seemed logical because they wanted to model the number of distressed parts rather than perfect flights.

Why does including launches with no failures help model failures?

The launches with no failures all had warmer launch temperatures. They are the data points at the bottom-right of the graph that anchor down that end of the trendline. Those launches provide crucial contextual information for the model about the conditions that produce no failures versus those that do. Warmer = safer. Colder = more dangerous.

Whenever you perform an analysis, you need to be aware of any missing data and its potential impact on the analysis.

Missing Data Conclusion

When handling missing data, it's crucial to understand the data gaps and their potential impact, whether you are a researcher conducting experiments, a data scientist working with large datasets, or an analyst evaluating others' work. Each type of missing data—whether Missing Completely at Random (MCAR), Missing at Random (MAR), or Missing Not at Random (MNAR)—demands a specific approach to preserve the integrity of research findings.

For MCAR, ignoring these gaps often suffices without compromising results. However, for MAR and MNAR, analysts must use more sophisticated techniques, such as multiple imputation or advanced modeling, to ensure that analyses remain unbiased and representative of the target population.

Ultimately, a thorough understanding of the data's structure and the missingness mechanism should guide the choice between accepting, deleting, or imputing missing data. Such meticulous consideration helps deliver robust, reliable, and actionable insights.

Aggregated Data

Data aggregation is a vital process involving collecting and summarizing data concisely. This method transforms rows representing individual observations—often sourced from diverse origins—into comprehensive totals or summary statistics. Aggregated data, typically housed in data warehouses, enhances analytical capabilities and significantly speeds up querying large datasets.

Data aggregation plays a vital role in statistical analysis and business intelligence. Organizations can analyze patterns and trends across groups by summarizing data, facilitating insightful business analyses. These summary numbers allow analysts to access vast amounts of data efficiently and reduce computational demands while enabling them to explore and interpret large datasets swiftly.

With the continuous expansion of organizational data, effective data management becomes essential. Data aggregation not only streamlines access to frequently used data but also enhances the efficiency of data retrieval. This process is particularly beneficial in producing aggregated measurements such as sums, averages, and counts, which are essential for in-depth business analysis.

Besides enhancing data accessibility and analysis, data aggregation can also serve a role similar to data anonymization. The aggregation process obscures personal details by merging individual records into group summaries, thereby protecting privacy. For example, summarizing employee salaries by department hides individual compensation data.

While data aggregation has impressive strengths, it also comes with inherent challenges. These include the potential loss of nuanced information, the inability to assess relationships between variables at the individual level, and the risk of misinterpreting aggregated outputs, which can lead to incorrect decisions—more on the limitations shortly.

Examples

Data aggregation is invaluable for many fields, including finance, business strategy, product planning, pricing strategies, operations optimization, and marketing. It equips data analysts, scientists, warehouse administrators, and industry experts with the tools to derive actionable insights from complex datasets. This process is instrumental across various domains, enabling organizations to assess vast amounts of information quickly and efficiently.

Healthcare Utilization: Hospitals and clinics can aggregate patient visit data to determine the most common ailments each season. This aggregated data helps in resource planning and public health advisories.

Retail Sales Trends: Retailers can use data aggregation to track average sales volumes per category, identifying trends and seasonal spikes in consumer purchasing behavior. This type of analysis enables better stock management and marketing strategies.

Traffic Management: City planners often use aggregated data from traffic sensors to analyze peak traffic flow times and congestion points. This information is crucial for traffic light scheduling and urban planning to improve city traffic conditions.

Energy Consumption Patterns: Utility companies aggregate data on energy usage to understand consumption patterns across different regions. This aggregated information is vital for forecasting demand and managing energy supply efficiently.

Educational Performance Metrics: School districts can aggregate student performance data to identify trends in academic achievements and areas needing attention. This data helps tailor educational programs and interventions to boost overall student performance.

Limitations

While data aggregation is invaluable for simplifying complex data sets and enhancing data analysis, it has limitations. Recognizing these shortcomings is essential for organizations to use aggregated data effectively and avoid potential pitfalls.

The Ecological Fallacy: Misinterpretations at the Individual Level

A critical limitation of using aggregated data is its inability to accurately represent relationships between variables at the individual level, a concept known as the ecological fallacy. This fallacy occurs when analysts incorrectly attribute relationships in aggregated data to individuals. The problems arise because the data aggregation process produces a significant loss of information, which can obscure or reverse relationships between variables at the individual level.

For example, suppose analysis of aggregated data suggests that regions with higher education levels also have higher incomes. In that case, one might wrongly infer that higher education directly increases income for individuals in those regions. However, this overlooks the possibility of significant variability within each area, such as high earners with varying education levels, thus misrepresenting the actual economic dynamics and possibly leading to flawed decisions.

In short, statistical analysis of aggregated data can be invalid if you apply the results to individuals. If you need analysis at the individual level (e.g., the effect of education on income for individuals), use individual-level data rather than aggregated data.

Loss of Detail and Variability

One of the primary drawbacks of data aggregation is the loss of granular details. Summarizing data obscures nuances and individual variations, which might be critical for certain types of analysis. For businesses, this loss of detail can lead to oversimplified conclusions that might not reflect complex market dynamics or consumer behaviors.

For instance, if an education study reports average test scores by district, it might overlook each district's wide range of scores. A district might show a high average score, concealing that while many students perform exceptionally well, a significant number might struggle severely. This loss of detail can prevent educators and policymakers from recognizing and addressing such disparities.

Aggregated data, while applicable for broad analysis, can sometimes lead to misleading interpretations if not carefully handled. This problem occurs because aggregation typically involves averaging or summing up values, which can mask significant outliers and anomalies. As a result, decision-makers might make strategies without considering the variation within groups.

Understanding data variability is the core theme for Chapter 7.

Data Integrity Issues

Aggregating data also involves potential data integrity risks. Errors in the initial data collection, processing stages, or managing missing data can magnify during data aggregation. Furthermore, inconsistencies in how data are aggregated across different sources or times can lead to discrepancies that are hard to detect and rectify.

Time Sensitivity

Another challenge with data aggregation is related to its relevance over time. Aggregated data can quickly become outdated in rapidly changing environments, such as financial markets or consumer trends, where real-time data is crucial. Relying solely on aggregated data can prevent organizations from responding swiftly to market changes.

Simpson's Paradox

A particularly intriguing issue related to data aggregation is Simpson's Paradox, where trends apparent in several different groups disappear or reverse when these groups are combined. This paradox highlights the importance of examining data at both the aggregated and disaggregated levels to avoid erroneous conclusions based on aggregated data alone.

Simpson's Paradox occurs when you combine subgroups into one group. Aggregating data can change the apparent direction and strength of the relationship between two variables. Imagine examining a set of data that tells one story when viewed as a whole, but a completely different narrative emerges when you break it down into subgroups. This paradox challenges our intuition and underscores the critical importance of a nuanced approach to data analysis.

In this section, we will explore the concept of Simpson's Paradox through real-life examples, including the infamous UC Berkeley gender admissions case and recent COVID-19 vaccination data, demonstrating how essential it is to consider all variables before drawing conclusions.

Simpson's Paradox occurs because a third variable can affect the relationship between a pair of variables. These are the confounding variables we covered in the previous chapter. To understand the correct relationship between two variables, you must factor in the influence of confounders.

The problem occurs when the process of aggregating data excludes confounding variables. Usually, this happens unintentionally. The researchers might not realize the consequences of their actions.

How It Works Graphically

Let's look at Simpson's paradox graphically before returning to the admissions example. The data below show one group with a negative correlation between the X and Y variables. As X increases, Y tends to decrease.

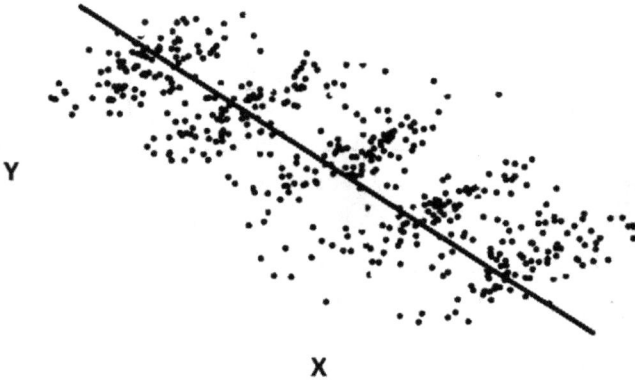

Now we'll factor in the subgroups. Below, it's easy to see how there is actually a positive relationship between X and Y within the subgroups. Combining the data and ignoring the subgroups obscured that relationship. That's how Simpson's paradox distorts the results.

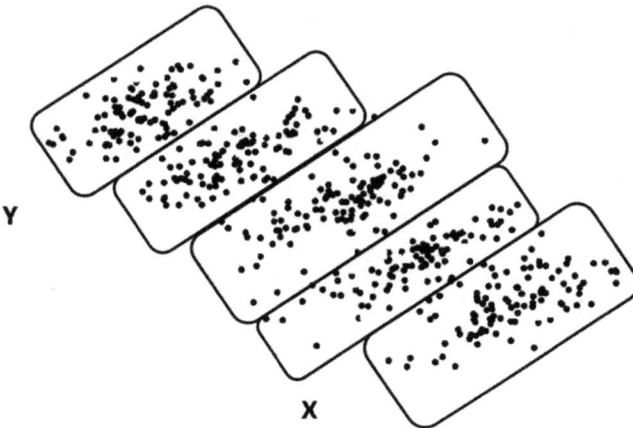

In the context of Simpson's Paradox, the subgroups capture the confounding variable. By aggregating the data, you are effectively removing the confounder from the analysis, and it distorts the results.

Example: UC Berkeley Admissions 1973

If you want to compare the admissions rates for men and women at UC Berkeley, it seems logical that you can look at the overall rates. Below are the 1973 acceptance rates:

Men	Women
45%	30%

It sure appears that Berkeley prefers men and disadvantages women. However, thanks to the Simpson's Paradox, there is more to the story.

Unfortunately, aggregating the data from all departments removes departmental differences from the analysis. Some departments have much lower acceptance rates than others. They're more selective. The following two factors create the misleading, unbalanced acceptance rates in the previous table:

- Women tended to apply for the harder departments, lowering their overall acceptance rate.
- Men were inclined to apply for the easier departments, boosting their rates.

To determine whether the selection process favors men, we need to assess the data at the departmental level and compare acceptance rates within each department. This method counters Simpson's paradox by accounting for each department's acceptance rate, allowing for valid comparisons.

Let's look at the data! There are 85 departments. The table shows the largest six.

Department	Men	Women
1	62%	82%
2	63%	68%
3	37%	34%
4	33%	35%
5	28%	24%
6	6%	7%

Comparing the rates within departments paints a different picture. Women have a slight advantage over men in most departments.

The subgroup analysis accounts for the confounding variable of the varying admission rates.

Example: COVID Death Rates

Simpson's Paradox occurs in numerous contexts. During the pandemic, analysts observed it in media reports of more COVID-19 deaths among the vaccinated than the unvaccinated. In September 2022, 12,593 COVID-19 deaths occurred in the United States. Of those, 39% were unvaccinated, while 61% were vaccinated.

We covered this in Chapter 2 with the base rate fallacy where we found that the unvaccinated were 2.5 times more likely to die. Here, we'll incorporate additional relevant factors to fine-tune the result.

It turns out that the relationship between being vaccinated and having a higher percentage of deaths is a fiction created by aggregating data and tossing out relevant information—Simpson's Paradox.

In the United States, the COVID-19 vaccinated population tends to be older and has more risk factors than the unvaccinated. This group naturally tends to have worse COVID-19 outcomes. It's analogous to the UC Berkeley example where women tended to apply to the more selective departments. In both cases, the two groups are dissimilar in ways that affect the outcome.

In the COVID example, vaccinated people tended to have more health risks. When you adjust for age and other risk factors, the CDC finds that COVID vaccinated and boosted individuals have an 18.6 times *lower* risk of dying from COVID-19. The vaccines are working!

To wrap up, Simpson's Paradox occurs when you fail to account for relevant information when analyzing data. This paradox occurs when you aggregate data and lose essential details. With the enrollment example, you get opposite results when you look at the overall acceptance rates by gender but don't consider the varying departmental acceptance rates. For the COVID example, you get distorted results when you assess the overall death COVID percentages by vaccination status without accounting for underlying risk factors.

Caution!

It shouldn't be surprising that discounting relevant factors will distort your results. But it is shocking how easily it can happen if you don't watch for it!

Simpson's Paradox is a powerful reminder of the complexities inherent in data analysis. As we've seen through examples from university admissions and public health, failing to account for subgroup variations can lead to conclusions that are not only incorrect but potentially misleading. This paradox teaches us the importance of vigilance and precision in statistical analysis, urging researchers to delve deeper into the data rather than accepting surface-level insights.

By understanding and acknowledging the impact of confounding variables and the dangers of data aggregation, we can prevent misinterpretations and ensure that our analyses truly reflect the reality they intend to capture. Be sure to do the following:

- Always question the data.
- Look beyond the aggregates.
- Strive for clarity and accuracy in every dataset you encounter.

By doing this, you can ensure that your analytical results accurately reflect the underlying trends and patterns in the data.

Aggregated data is hugely important in data warehouses and the big data context. It allows for efficient access to large amounts of data. However, be aware of its limitations. It can't capture individual-level relationships, and it can obscure those pesky confounding variables!

Cautions About Graphing

I love graphs! They are excellent choices for a clear, intuitive view of the data. However, I have two cautions about them.

Graphs are subject to manipulation. The manipulation might even have good motivations, such as making it look more dramatic. By changing aspects like the axes or bin sizes on histograms, you can substantially change the appearance of graphs. The same data can look very different and convey different interpretations.

For example, the relationships in the following two scatterplots look different, but they display the same data. The first graph is more dramatic.

Scatterplot of Weight kg vs Height M

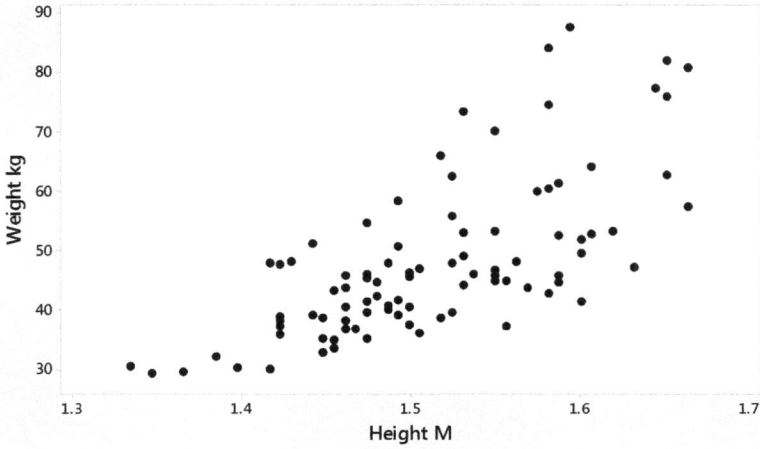

Scatterplot of Weight kg vs Height M

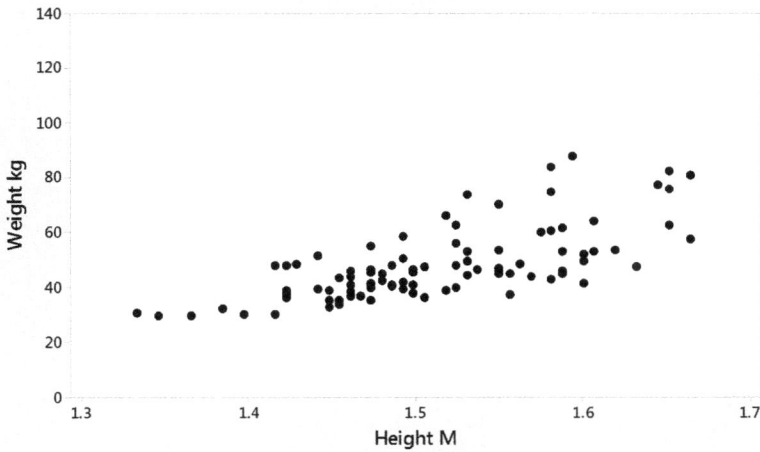

The following two boxplots also display the same dataset. The difference between the two groups appears more significant in the first plot.

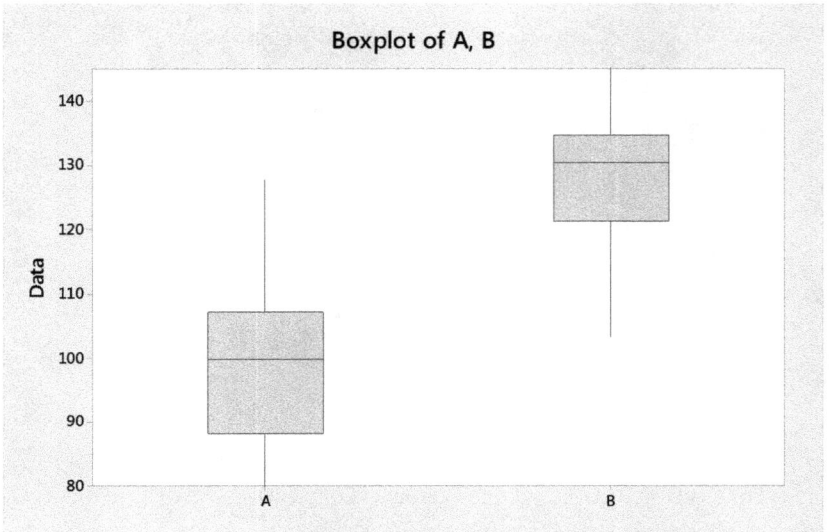

Boxplot of A, B

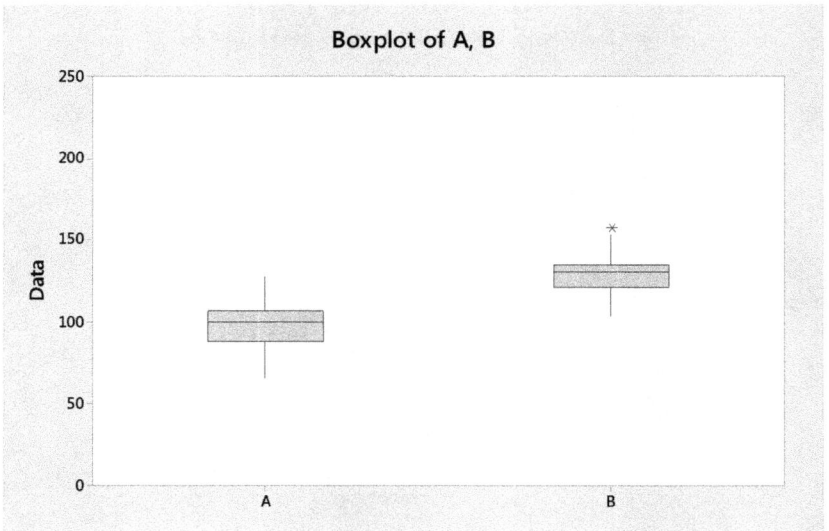

Boxplot of A, B

Anytime you look at a graph, carefully evaluate the scales!

Numeric measures are more objective and harder to manipulate. Correlation coefficients and the difference between means don't change based on the graph's axes! That's why I often combine summary statistics and graphs.

Graphs can help portray relationships in the data. The previous graphs displayed correlations between variables and differences between groups. Graphs are excellent tools for bringing all these relationships to life visually. However, if your goal is to use the sample to estimate relationships in a population, you'll need to use a hypothesis test.

In broad strokes, these tests determine whether relationships in your sample data are likely to exist in the population. The patterns you see in the graphs might be flukes due to random sampling error rather than denoting real relationships in the population. After all, if you take some random variation between groups and throw in some automatic graph scaling designed to make striking graphs, you can get impressive looking results.

Hypothesis tests are critical in separating the signal (real effects in the population) from the noise (random sampling error). This protective function helps prevent you from mistaking random error for a real effect.

Hypothesis Testing Overview

In this chapter, we've covered sample size, outliers, missing data, aggregated data, and graphing. I don't know about you, but I don't think any of those topics are exceptionally sexy. However, I hope you see how crucial they are. Being an analytical thinker requires you to dig into the details. Any one of them can change the entire nature of your results.

And now we're moving on to p-hacking, which does sound kind of sexy. The term's inventors coined it intentionally to attract attention because a problem exists in academic studies and probably across all data analytic scenarios—albeit less recognized.

P-hacking is a set of methodology choices like those described in this chapter and they relate to p-values and statistical significance. Consequently, a short primer is needed before getting to p-hacking.

Hypothesis testing is a form of inferential statistics where you use a sample to learn about populations. When you draw a random sample from a population, these tests help you determine whether your sample provides strong enough evidence to conclude that an effect or relationship in the sample also exists in the population. There's always the chance that luck of the draw (i.e., random sampling error) produced the interesting sample results.

Hypothesis tests use p-values and significance levels to make this determination.

- P-values measure the strength of the evidence in your sample that an effect exists.
- The significance level indicates how strong the evidence must be to conclude that an effect exists in the population. It is also the test's false positive error rate.

When your p-value is less than or equal to your significance level, your results are statistically significant. You can conclude that the effect you see in your sample also exists in the population.

These conclusions are crucial in science and elsewhere when you want to generalize your results to others. Again, when scientists study the effectiveness of a new drug, they don't want to learn only whether it works in the sample. They'd like to know that it is effective for the whole population. Enter hypothesis testing!

In this manner, hypothesis testing is a protective measure that helps prevent you from falsely claiming that random noise in your sample is a population effect. That would be a false positive or false discovery.

The significance level also quantifies the probability of that happening.

For instance, all studies using the standard significance level of 0.05 have a 5% chance of indicating an effect exists in the population when it doesn't. There's no way to get that down to 0%, but you can keep the false positive rate very low.

Statisticians refer to that false positive rate as the Type I error rate, which equals the significance level. However, that's only true when analysts follow all the proper methodologies. Improper choices can artificially create statistically significant studies, dramatically inflating the Type I error rate. That's p-hacking.

P-Hacking

P-hacking is a set of statistical decisions and methodology choices during research that artificially produce statistically significant results. These decisions increase the probability of false positives, where the study indicates an effect exists when it actually does not. P-hacking is the manipulation of data analysis until it produces statistically significant results, compromising the truthfulness of the findings. This problematic practice undermines the integrity of scientific research.

In academia, p-hacking occurs because high-impact journals strongly favor statistically significant results. For researchers, publishing in these prestigious outlets is a career-boosting achievement. However, this prestige comes with pressure that can tempt researchers towards the perilous path of p-hacking.

The term p-hacking was born during a replicability crisis within the scientific community. Scientists were struggling as some of their landmark findings were failing to replicate. A large-scale research project repeated 100 previously significant psychology studies and found that a shocking 64% were not significant the second time (Open Science Collaboration, 2015). The colossal failure to replicate two-thirds of the significant results suggests that most of the original studies were false positives.

With growing unease, investigators tried to find the causes of the false positives. The suspect? Some deeply ingrained research practices. As the plot thickened, it became clear that p-hacking was a major player in this unfolding replication crisis.

All studies have a false positive rate. That's the probability of concluding an effect or relationship exists (i.e., significant results) when it does not. Statisticians refer to that as the Type I error rate in hypothesis testing. When you do everything correctly, the error rate equals your significance level (e.g., 0.05 or 5%).

P-hacking jacks up that false positive rate, sometimes drastically! False positive studies tend not to reproduce significant results when scientists repeat them—explaining the replication crisis. Clearly, the 64% reproducibility failure in the aforementioned study is *much* greater than expected!

The takeaway is that p-hacking's detrimental effects are real and not theoretical. Scientists have already noticed its impact in the literature with the replication crisis. Additional studies have found an overabundance of p-values just below 0.05. These results suggest that researchers have tweaked their studies until their p-values are just below the standard significance threshold of 0.05 (Simonsohn, Nelson, & Simmons, 2013).

Origin and Debate around the Term

Simonsohn, Nelson, and Simmons are authors of landmark studies in p-hacking and introduced the term at a psychology conference in 2012. They wanted to create a memorable name for the set of practices. While catchy, the term has sparked a debate among statisticians. Critics argue that the word "hacking" implies it refers only to intentionally deceptive manipulation. In actuality, the word applies to both unintentional and intentional cases.

Unintentional P-Hacking: Many researchers don't fully realize they're p-hacking. With so many ways to analyze data (imagine wandering through a maze of paths), it's easy to veer toward biased decisions unknowingly. It's like convincing ourselves that the shorter, easier route is the 'right' one, even when it's not. Cognitive biases, such as confirmation bias, play a role here.

Intentional P-Hacking: P-hacking can be willful manipulation using an iterative trial-and-error method that homes in on significant results. Here, researchers knowingly twist their analysis to create desired outcomes. It's akin to deliberately changing the evidence at a crime scene to create a misleading narrative.

Whether done knowingly or not, p-hacking clouds the truth and jeopardizes the pursuit of knowledge. Let's examine how p-hacking occurs and the best practices for avoiding it.

Methods

P-hacking covers a wide range of methods. The difficulty is that all studies require researchers to make numerous decisions about data collection, variable manipulation, analysis techniques, and reporting the results. Making the correct choices is crucial to producing valid results.

When done correctly, some of the following methods can be legitimate decisions. P-hacking refers to cases where researchers make poor choices that produce unwarranted statistically significant results.

Research by Stefan and Schönbrodt (2023) has identified the following most common p-hacking methods. However, there are numerous others. As you read through these examples, remember that researchers have reviewed previous research and identified these issues.

Again, it's not theoretical. It has been observed. It makes you wonder how often these issues occurs in less transparent data analysis realms where there's often great pressure to find predictive models?

Stopping Rules

P-hackers might stop collecting data once they achieve a significant result, ignoring the need for a predetermined sample size.

Optional stopping, or as some call it, 'data peeking,' is when a researcher keeps testing their hypothesis as they gather data. The moment they hit a significant result, they stop collecting data. It's like prematurely declaring victory in a game before it's officially over—when your team first pulls ahead.

This premature termination of data collection is a p-hacking technique that inflates Type I errors.

Outliers

Outliers can significantly impact your data. P-hackers might choose to remove outliers based on whether it helps them achieve significance. The decision about outliers ideally should be based on theoretical grounds about the variable and measurement issues relating to specific observations. The outliers' impact on the p-value should not be a factor at all.

'Data trimming' is a common form of p-hacking where researchers selectively exclude outliers. There's room for bending the rules with so many ways to identify outliers (39 common methods!). Plus, reporting on how researchers handle outliers is often sketchy, making it easier to hide data trimming. Some studies even fail to mention it, leading to unexplained differences in sample sizes and degrees of freedom. So, watch out for the elusive outliers!

Removing outliers reduces the data's variability, thereby increasing statistical power. However, you're potentially removing legitimate data points and basing your analysis on an unnatural dataset.

Variable Manipulation

Researchers frequently need to manipulate their variables for various legitimate reasons. But p-hackers make changes to produce statistical significance.

In the p-hacking realm, analysts might slice, subgroup, or subset their data in ways that produce significance when the original arrangement does not. For example, combining comparison groups, recoding a continuous variable into a discrete variable, and looking at only a subset of the sample can produce statistical significance that wouldn't exist otherwise. In regression analysis, unnecessarily transforming the independent and dependent variables can create unwarranted significance.

According to Stefan and Schönbrodt, one of the most common forms of p-hacking is changing the primary outcome variable *while* the study is ongoing. Researchers peek at the data and then change their primary outcome to a variable that seems more likely to achieve significance. For instance, a diabetes medicine study starts by tracking blood glucose levels but changes to another outcome measure after six months because it is more likely to produce statistically significant results.

This 'moving of the goalposts' is a classic example of p-hacking, altering the study's outcome variable mid-stream to achieve statistical significance.

In the worst cases, researchers use a trial-and-error approach of manipulating variables until it produces statistical significance. Changing the study design and analysis to chase significance inflates the false positive rate.

Excessive Hypothesis Testing & Multiple Comparisons

When researchers perform many hypothesis tests, they increase the likelihood of stumbling upon a statistically significant result purely by chance. A single hypothesis test has a false positive (Type I error rate) equal to the significance level (e.g., 0.05). For a set of hypothesis tests, the family error rate increases for each additional test.

It's like flipping a coin many times; sooner or later, you will get a string of heads. But remember, this doesn't imply that the coin is biased, just as a significant result amid numerous tests doesn't necessarily signify a meaningful finding.

Similarly, the more groups researchers compare, the higher the chances of finding a significant result purely by chance. Correcting for multiple comparisons is essential to maintain the integrity of the results.

Additionally, p-hackers might run multiple variations of the same analysis, try similar analyses, relax assumptions, and alter little things each time—like the control variables or subsets of data used. They continue this process until they stumble upon a significant result.

Researchers need to limit the testing they perform during a study and use the proper corrections for multiple comparisons and hypothesis tests.

Excessive Model Fitting

This problem is similar to excessive hypothesis testing but relates to fitting many different regression models. P-hackers can experiment with numerous statistical models until they find one that delivers the desired results. This process becomes problematic when model selection is driven by statistical significance rather than the appropriateness for the data and research question.

While it's essential to control for confounding variables by including them in the model, it can be a double-edged sword. Deciding which variables to control can be twisted into another form of p-hacking, especially if researchers base the decision on chasing statistical significance rather than for theoretical and subject-area reasons.

If you fit many models and use statistical significance to guide you, you can produce models that "explain" relationships in randomly generated data.

This sneaky p-hacking technique isn't limited to regression analyses; it can happen anytime there's an option to pick and choose variables. In Chapter 8, I cover this problem in more detail.

Selective Reporting of Results

This p-hacking method involves cherry-picking the outcomes and hypothesis tests for reporting while failing to discuss nonsignificant results and changes in the study design. This method creates a false impression of the results' strength by overemphasizing the significance and downplaying the weaknesses and nonsignificant findings.

For example, a study might measure many different outcomes and find a significant result for only one. Or they conduct many hypothesis tests and only present the few that give statistically significant results, conveniently leaving out the ones that don't. This approach is akin to showing a highlight reel without the unimpressive plays.

If a study mainly finds nonsignificant results, you'd have good reason to question its few significant findings. However, if the reporting doesn't discuss the slew of nonsignificant conclusions, you won't know the proper context for evaluating the results.

Additionally, a series of nonsignificant findings followed by a significant result is a red flag for the previous trial-and-error methods I describe above.

Best Practices

As we delve into p-hacking methods, it becomes increasingly clear how easy it is to veer into these practices, intentionally or not. It underlines the importance of sound analytical training and an unwavering commitment to integrity. Aim to tell the story of the data as it is, not as we'd like it to be.

P-hacking can quietly erode the foundations of scientific research. But don't despair. Here are some best practices to keep you on the right path.

Develop a Clear Research Plan

Create a detailed plan before conducting the research. It should include your hypotheses, data collection methods, and analyses. This clear roadmap helps prevent you from going down the p-hacking trial-and-error approach of performing variable manipulation and data analysis variations until you get significant results.

Preregister Your Studies

Publicly specify your research plan before conducting the study. This approach further reduces the temptation to deviate based on interim findings. And it signals other researchers that they can take your research more seriously. You can preregister studies at places like the Center for Open Science (cos.io).

Transparent Reporting

Report all your steps, even the not-so-successful ones. Honesty is your best ally in research. This transparency includes defining comparison groups in advance and reporting all variables, conditions, data exclusions, tests, and measures.

Education and Training

Many p-hacked studies stem from not understanding the pitfalls rather than malicious deception. Ensure you have a strong understanding of statistical principles and maintain an awareness of the pitfalls of p-hacking. Continuous learning is an essential tool in any researcher's kit. It's one of the many reasons why understanding statistics is vital.

Ultimately, remember that every decision made during statistical analysis affects the results. P-hacking might not always be a deliberate act of deception. It can often stem from a lack of understanding of statistical principles.

Adhering to these best practices can keep our research robust and our findings credible. Avoiding p-hacking isn't just about securing valid results; it's about preserving the integrity of the scientific process.

Stay ethical and keep crunching those numbers responsibly!

Closing Thoughts

Throughout this chapter, you saw how the nuts and bolts of data analysis can dramatically affect the results. The details matter. And there are a ton of details. Analysts can accidentally or intentionally manipulate those particulars to produce results that produce results that benefit themselves.

This issue directly ties into all those cognitive biases I covered in Chapter 1. In data analysis, these biases can significantly skew the methodology and outcomes of studies, often subconsciously influencing analysts' decisions in ways that can compromise the integrity and validity of the results.

Analysts need to make many decisions during studies, and many of those decisions haven't been preplanned. So, they must decide on the spot and will often have an inclination of how the decisions affect the results. That opens the door for cognitive errors to come into play.

For instance, confirmation bias might lead an analyst to favor data or testing methods that affirm their pre-existing beliefs or hypotheses, potentially overlooking contradictory evidence. Similarly, the self-serving bias could drive analysts to select or interpret data in a way that reflects favorably on their performance or benefits their employer rather than objectively assessing the information. Anchoring bias is another peril, where analysts might give disproportionate weight to the first piece of information they encounter—such as an initial study or a prominent dataset—thereby shaping all subsequent analysis around this anchor, despite new data suggesting alternative narratives.

Each of these biases can lead to methodological decisions that inadvertently mold the analyses to fit the expected or desired outcomes rather than letting the data shape its own story, ultimately risking the accuracy and reliability of the conclusions drawn. Imagine cases where analysts try various types of analyses or excessively play with different regression models until one seems to be better because it fits with their preconceived notions.

It's not hard to imagine that NASA's desire to stick with the launch schedule for financial reasons could have influenced the space shuttle *Challenger* safety assessment—even if only subconsciously. And when the results agreed with their hopes for launching, there was less reason to double-check them.

Always examine your analytical processes and results carefully—*especially* when the results agree with what you were hoping or expecting them to say!

Statisticians recommend you plan out as much as possible before collecting and analyzing the data. If you plan these decisions before seeing how they affect the results, you're more likely to make the correct choices. Correct in this context is defined as making decisions that allow the analysis to produce results that match reality—and might not match your expectations or wishes.

We've developed these slick analytical procedures to overcome our innate mental limitations, yet we need to be on guard against cognitive errors even while using these tools. And, even if they're not secretly driving the choices, be aware that how you handle outliers, missing data, and aggregated data is vital for producing accurate results.

✿

CHAPTER 7

Analytics: Variability and Signal vs. Noise

In this chapter, we'll continue assessing fundamental aspects of statistical analysis. Statisticians have considered these issues for over a century and incorporated them into their analytical procedures.

At the conceptual level, variability is a simple concept. If you take a random sample of people and measure their heights, you expect them to have different heights. Some will be taller than others.

So, you expect that. Now, suppose you draw a random sample of men and women and try to determine which gender is taller on average. Pretend you're an alien and don't know! Because height varies randomly from person to person, these aliens might select several women purely by chance who are taller than the men they measure.

In this example, the aliens did everything correctly in their random sampling and measurements. It was just bad luck. That happens when working with random samples. In this manner, variability makes it more difficult to draw accurate conclusions by making the picture fuzzy.

In the previous chapter, we briefly discussed the problems of over-looking variability when discussing aggregated data. In this chapter, I expand on how variability affects your ability to draw conclusions from data. A common analogy is that you need to separate the signal (the effect) from the noise (variability).

Let's dig deeper!

When you think about statistical analyses, you tend to think about the measures of central tendency. It's often the mean, but it can also be the median or mode.

- What is the group mean?
- What is the mean difference between the treatment and control group?
- What's the mean outcome?

And so on.

However, so much of statistics involves factoring in variability. The variability in a population or dataset is at least as significant as the central tendency. It can cause you to miss real effects that exist and create the appearance of effects where they don't exist.

Let's take a step back and first get a handle on why understanding variability is so essential.

Analysts frequently use the mean to summarize the center of a population or a process. While the mean is relevant, people often react to variability even more. When a distribution has lower variability, the values in a dataset are more consistent. Conversely, when variability is higher, the data points are more dissimilar, and extreme values become more likely. Consequently, recognizing the variability in data helps you understand the likelihood of unusual events.

In many situations, extreme values can cause problems. Have you seen a weather report where the meteorologist shows extreme heat and drought in one area and flooding in another? It would be nice to average those together! Frequently, we feel discomfort at the extremes more than the mean. Understanding variability around the mean provides critical information.

Variability is everywhere. Your commute time to work varies a bit every day. When you order a favorite dish at a restaurant repeatedly, it is different each time. The parts that come off an assembly line appear identical but have subtly different lengths and widths.

These are all examples of real-life variability. Some degree of variation is unavoidable. However, too much inconsistency can cause problems. If your morning commute takes much longer than the mean travel time, you will be late for work. If the restaurant dish is much different than usual, you might not like it. And, if a manufactured part is too much out of spec, it won't function.

Some variation is inevitable, but problems occur at the extremes. Distributions with greater variability produce unusually large and small values more frequently than distributions with less variability.

Statisticians measure variability in several ways, but the standard deviation is the most common.

The standard deviation (S.D.) is a single number that summarizes the variability in a dataset. It represents the typical distance between each data point and the mean. Smaller values indicate that the data points cluster closer to the mean—the values in the dataset are relatively consistent. Conversely, higher values signify that the values spread further from the mean. Data values become more dissimilar, and extreme values become more likely.

The standard deviation uses the original data units, simplifying the interpretation. For this reason, it is the most widely used measure of variability.

Let's take a look at two hypothetical pizza restaurants. They both advertise a mean delivery time of 20 minutes. When we're ravenous, they both sound equally good. However, this equivalence can be deceptive. We should analyze their variability to determine which restaurant to order from when we're hungry.

Suppose we study their delivery times, calculate the variability, and find that their variabilities are different. We've computed the standard deviations for both restaurants. One place has a S.D. of 10 minutes while the other's is 5. How significant is this difference in getting pizza to their customers promptly?

The following graphs display the distribution of delivery times and provide the answer. The restaurant with more variable delivery times has a broader distribution curve. I've used the same scales in both graphs so you can visually compare the two distributions.

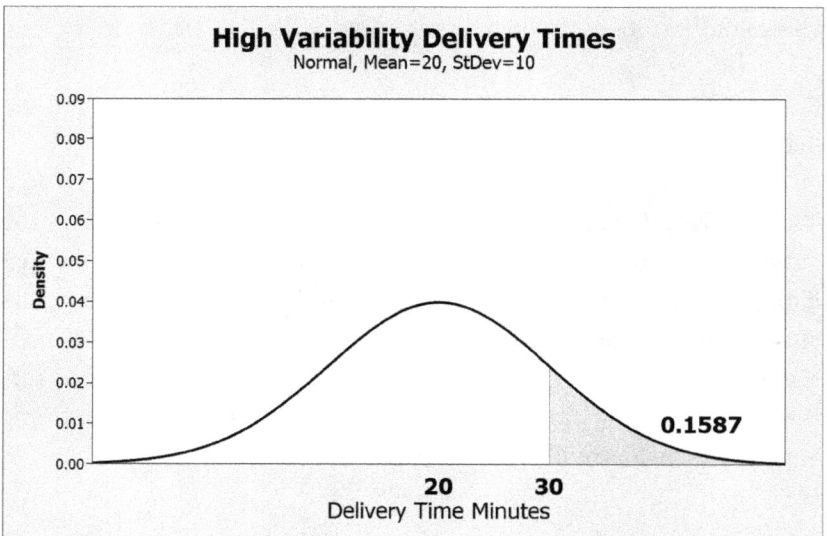

High Variability Delivery Times
Normal, Mean=20, StDev=10

0.1587

Low Variability Delivery Times
Normal, Mean=20, StDev=5

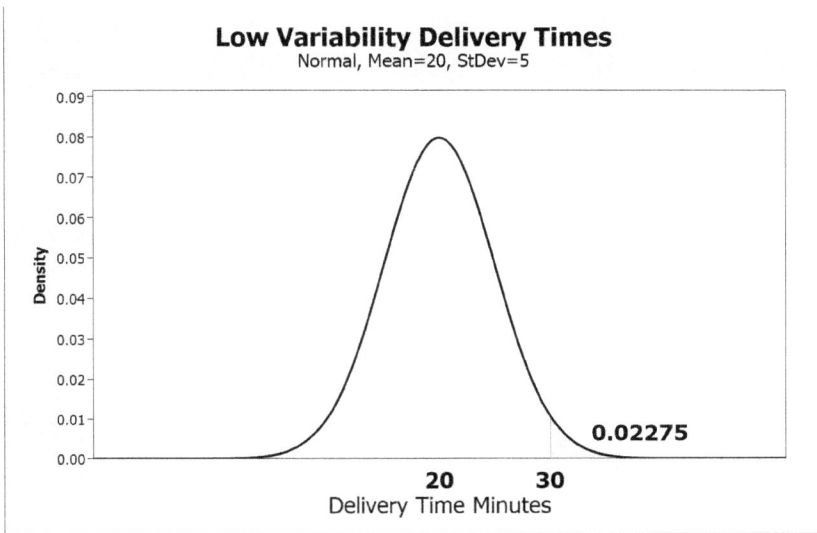

In these graphs, we consider a 30-minute wait or longer unacceptable. We're hungry after all! The shaded area in each chart represents the proportion of delivery times that surpass 30 minutes. Nearly 16% of the deliveries for the high-variability restaurant exceed 30 minutes. On the other hand, only 2% of the deliveries take too long with the low-variability restaurant. They both have an average delivery time of 20 minutes, but I know where to place my order when hungry!

As this example shows, the central tendency doesn't provide complete information. We also need to understand the variability around the middle of the distribution to get the full picture.

Example: Jumping Impacts for Bone Density Study

As you saw above, focusing solely on the mean can lead to misleading conclusions. Let's look at a real example that I faced. The following example comes from the bone density study I mentioned earlier.

Our research examined bone density among teenage girls by evaluating the impact of jumping from 24-inch platforms 30 times every other school day to see if this activity would enhance their bone density compared to a control group. We aimed for each jump to produce a ground reaction force (GRF) of six times the jumper's body weight (BW).

These impacts were the treatment dose for our study. Hence, it's as crucial to quantify this form of treatment as a medication dosage in a clinical trial.

In a pilot study, each participant jumped onto a force plate five times to measure the impacts. The results were encouraging, showing an average impact of 6.13 BWs. However, we also observed high variability (standard deviation = 1.08 BWs).

The probability distribution plot of impact forces reveals that we could expect only 55% of participants to achieve impacts above 6 BWs, our target.

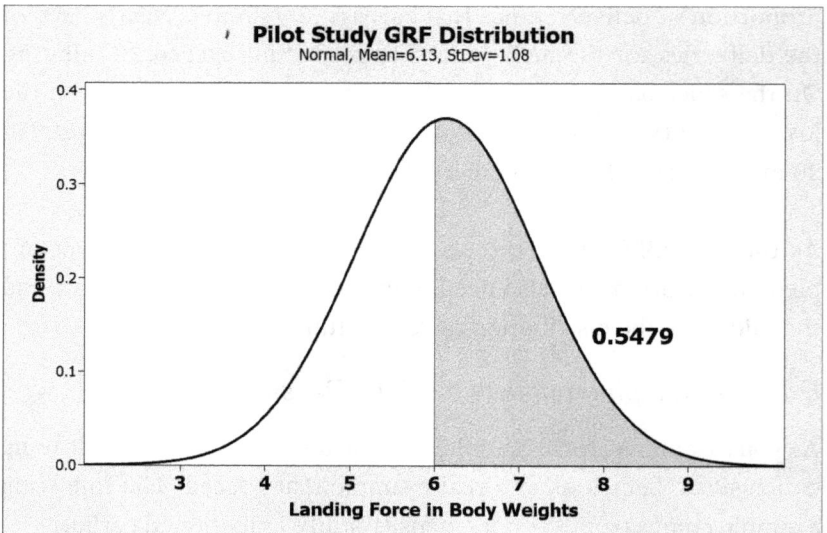

Unfortunately, while the mean impact is satisfactory, only half the subjects achieved it! I'll return to this problem later in this chapter so you can see how we resolved it.

Understanding variability is essential. Now, let's see how it affects other aspects of data analysis.

Sample Size: The Foundation of Reliable Statistics

Sample size is the number of observations or data points collected in a study. It is a crucial element in any statistical analysis because it is the foundation for drawing inferences and conclusions about a larger population. A study's credibility frequently rests on its sample size.
In Chapter 3, I discussed why you should care about *how* your sample was collected, but why should you care about its size? The answer ties directly into variability.

Imagine you're tasting a new brand of cookies. Randomly sampling just one cookie might not give you a true sense of the overall flavor—what if you picked the only burnt one? Trying more cookies gives you a more reliable perception of the cookies overall.

If all cookies in a population are identical (zero variability), you only need to sample one cookie to know what the average cookie is like for the entire population. However, suppose there's a little variability because some cookies are cooked perfectly while others are overcooked. You'll need a larger sample size to understand the proportions of perfect and overcooked cookies.

Now, let's complicate things a bit. Imagine the cookies have an entire range of over- and undercookedness. Additionally, some use sweeter chocolate chips than others. This increased variability requires an even larger sample to understand and define what an average cookie is really like.

The more variable they are, the more cookies you need to try to understand the full range of the cookies. Hmm. Lots of cookie tasting! And all in the name of science!

In this manner, the sample size determines how well your study represents the larger group. A larger sample size can mean the difference between a snapshot and a panorama, providing a clearer, more accurate picture of the reality you're studying.

Statistics bombard you. You can find them everywhere, such as in the news media, surveys, commercials, etc. Often, these statistics aren't meant to describe only the specific group of measured subjects. Instead, the goal is to infer properties about a larger population. This practice is called inferential statistics.

For example, when we read survey results, we are not learning about the opinions of only those who responded to the survey but about an entire population. Or, when we see averages, such as health measures and salaries, we're learning about them on the scale of a population, not just the few subjects in the study. Consequently, inferential statistics can provide extremely helpful information.

Inferential statistics is a powerful tool because it allows you to use a relatively small sample to learn about an entire population. However, to have any chance of obtaining good results, you must follow essential procedures that help your sample represent the population faithfully.

We covered those in previous chapters—sampling methodology, measurements, and an experimental design when you want to identify causal relationships. We also discussed the differences between sample statistics and population parameters.

In this chapter, we focus on how a sample statistic has a margin of error around it, representing the inherent uncertainty when you measure only a subset of the whole population. Conversely, a parameter is an exact but unknown value for the entire population.

Here's some shocking information for you—sample statistics are *always* wrong! When you use samples to estimate the properties of populations, you never obtain the correct values exactly.

Unfortunately, even when you diligently follow the proper methodology for performing a valid study, your estimates will almost always be at least a little wrong. I'm not referring to unscrupulous manipulation, mistakes, or methodology errors. I'm talking about cases where researchers use the correct sampling methodology and legitimate statistical calculations to produce the best possible estimates.

Why does this happen? Random sampling error is present in all samples. By sheer chance alone, your sample contains error that causes the statistics to be at least slightly off. Your data are not 100% representative of the population because they are not the entire population. Samples can never perfectly depict the population from which they are drawn.

All estimates are at least a little wrong, but they can be very wrong. Unfortunately, the media and other sources forget this point when they present statistics. Upon seeing an estimate, you should wonder—how significant is the difference between the estimate and the actual population value?

As discussed in Chapter 3, using a sample statistic to estimate a population parameter begins with using a representative sampling method because it generates unbiased results.

Recall the statistical definitions and curves for bias and precision. Representative samples are crucial for producing unbiased sample estimates. They're not systematically too high or low and are correct on average—their curves center on the proper value.

Now, we need to consider the precision of the sample estimates, which indicates how close the statistics are likely to be to the parameters. These calculations involve sample size and variability working together. But first, you need to understand sampling distributions to see how that works.

Sampling Distributions

So far, we've been looking at variability in the context of individual observations. However, we usually use samples to draw conclusions. You'll need to understand sampling distributions to understand how sample variability and sample size affect conclusions about a population.

A sampling distribution is a type of probability distribution created by drawing many random samples of a given size from the same population. These distributions help you understand how a sample statistic varies from sample to sample.

Sampling distributions are essential for inferential statistics because they allow you to understand a specific sample statistic in the broader context of other possible values. Crucially, they let you calculate a margin of error around your sample statistic. How wrong is it likely to be?

They describe the distribution of values for all manner of sample statistics. While the sampling distribution of the mean is the most common type, they can characterize other statistics, such as the median, standard deviation, range, correlation, and test statistics in hypothesis tests. I focus on the mean in this section.

Here's a simple example to help you understand the concept.

Repeated Apple Samples

Imagine you draw a random sample of 10 apples. Then, you calculate the mean of that sample as 103 grams. That's one sample mean from one sample. However, you realize that if you were to draw another random sample, you'd obtain a different mean. A third sample would produce yet another mean. And so on.

With this in mind, you decide to collect 50 random samples from the same apple population. Each sample contains 10 apples, and you calculate the mean for each sample.

At this point, you have 50 sample means for apple weights. You plot these sample means in the histogram below to display your sampling distribution of the mean.

I used Excel to create this example. I had it randomly draw 50 samples with a sample size of 10 from a population with $\mu = 100$ and $\sigma = 15$.

You can obtain the Excel file on the resource webpage for this book, which I include in the Introduction. Try it yourself! Note that anytime you open the file, Excel draws new samples and your graph will look somewhat different than the one below.

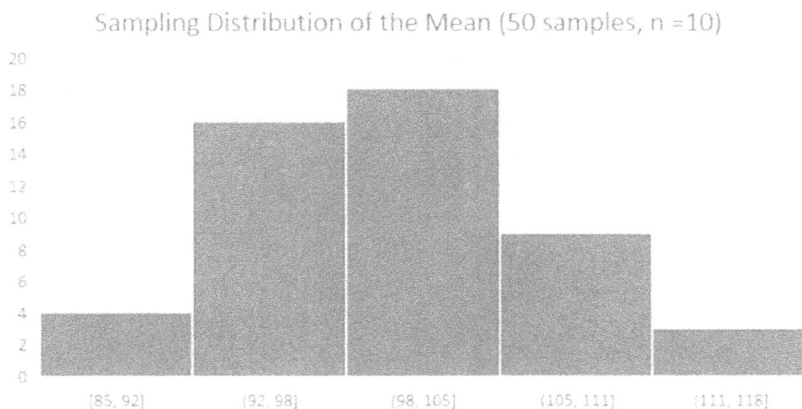

Sampling Distribution of the Mean (50 samples, n =10)

Jim Frost

The horizontal axis displays ranges of sample means. Remember, these are the means from samples with 10 apples in each, not individual values. The height of the bars indicates the frequency of sample means falling within each range.

Typically, you don't know the population values. Instead, you use samples to estimate them. However, we know the parameters for this simulation because I've set the population to follow a normal distribution with a mean (μ) weight of 100 grams and a standard deviation (σ) of 15 grams. Those are the parameters of the apple population from which we've been sampling.

This histogram shows us that our initial sample mean of 103 falls near the center of the sampling distribution. Means occur most frequently in this range—18 of the 50 samples (36%) fall within the middle bar. However, other samples from the same population have higher and lower means. The frequency of means is highest in the sampling distribution center near the population mean of 100 and tapers off in both directions. That's good because it indicates the results are unbiased or correct on average. None of our 50 sample means fall outside the range of 85-118. Consequently, it is unusual to obtain sample means outside this range.

Notice how the histogram centers on the population mean of 100, and sample means become rarer further away. It's also a reasonably symmetric distribution. Those are features of many sampling distributions. This distribution isn't especially smooth because 50 samples is a small number for this purpose, as you'll see.

There are several crucial points to notice before we move on.

While most sample means fall near the correct value, getting sample means further away is possible. Some of our 50 samples produced means between 85—92 and 111—118, even though the correct mean is 100. That's the luck of the draw producing fluky results that I mentioned. Sometimes, you can do everything correctly but still obtain misleading results.

The sample distribution has its own central tendency and variability. We'll explore those more when we discuss the bias and precision of sample estimates.

Sampling Distribution Properties

In the context of sampling distributions, a parent distribution refers to the distribution of individual items from which you're sampling, such as the apple population. When you understand the parent distribution, you can calculate the characteristics of the sampling distributions.

As you saw in the apple example, sampling distributions have a shape, central tendency, and variability.

When the parent distribution is normally distributed, its sampling distributions will also be normal and have specific properties for the central tendency and variability. Even nonnormal parent distributions (e.g., skewed) have sampling distributions that converge on a normal distribution with these properties as the sample size increases. This property is known as the central limit theorem.

	Mean	Standard Deviation
Parent Distribution	μ	σ
Sampling Distribution	μ	$\sigma/\sqrt{n}$

Where,
- μ and σ are the population parameters for the mean and standard deviation, respectively.
- n is the sample size.

Notice how the mean of the parent population is also the central value for the sampling distribution. That indicates the sample estimates are unbiased because they are correct on average. However, you must use a sampling method that produces a representative sample for this to occur. Other sampling methods, such as convenience sampling, tend to produce biased samples with sampling distributions centering on incorrect values—although that won't be obvious from the data itself.

However, the variabilities in the parent and sampling distribution are different. The variability for the parent distribution is a fixed value (σ), while for a sampling distribution it's related to σ but also depends on the sample size (n). From the formula, we know the variability for a parent distribution differs from its sampling distributions in all cases where n > 1. Additionally, each sampling distribution has a unique spread depending on its sample size.

Statisticians refer to the standard deviation for a sampling distribution as the standard error. Because we're assessing the mean, the variability of that distribution is the standard error of the mean. It represents the variability of sample means between random samples of the same size drawn from the same population.

The standard error formula is a ratio with the square root of the sample size in the denominator. This fact causes the value of the denominator to increase as the sample size increases. In turn, a larger denominator decreases the standard error. Consequently, sampling distributions based on larger sample sizes should have lower variability, causing them to cluster more tightly around the central value. That fact is crucial, as you'll see.

By understanding the impact of sample size on your results, you can make informed decisions about your research design and have more confidence in your findings. A large sample size can significantly enhance the precision of your study results.

Returning to Our Apple Simulation

We know what statistical theory and its equations say. Now, let's see how this works using random sampling to see how reality compares. We'll also get to see nice graphs!

We'll rerun our previous apple sampling simulation but on a massive scale. This time, I'll draw 500,000 samples instead of just 50. I'll use the Statistics101 simulation software to perform this simulation. I include links for this freeware and my script for it on the resource webpage for this book.

This simulation follows the same process as the Excel version. It draws random samples from a population with a mean of 100 and a standard deviation of 15. It calculates the sample means and plots them using a histogram. This setup is our previous simulation on steroids!

Each sample has a size of 10, and I'll add another simulation that quadruples the sample size to 40 to see what happens. Theoretically, quadrupling the sample size halves the standard error due to the square root in the standard error's denominator: $\sqrt{4} = 2$.

The sample sizes of 10 and 40 should affect the standard error of the mean but not the mean. Therefore, we'd expect both sampling distributions to center on $\mu = 100$, and that the standard error for n = 10 will be $15 / \sqrt{10} = 4.743$, and n = 40 should be half that: $15 / \sqrt{40} = 2.372$.

Let's run a simulation! Because this simulation draws so many more samples, it produces a smooth distribution curve that reveals the underlying function. This graph displays the distribution of sample means rather than individual values. It's a histogram like the one in Excel but with many more samples and tiny bars!

I'll have the simulation software calculate the mean and standard error of the sample means, which should be close to the theoretical values for both sample sizes. Please remember that the two curves represent the sample means, not the individual observations.

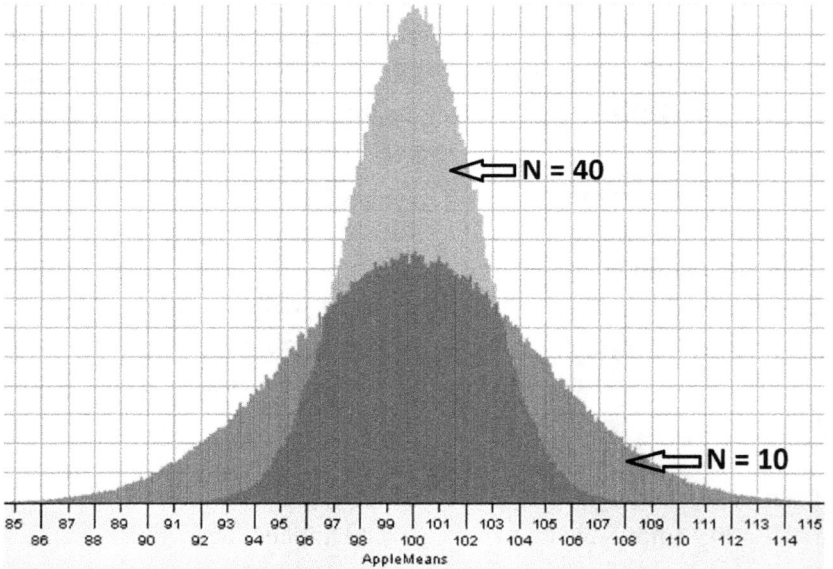

```
Mean10:  100.00070541860869
StandardError10:  4.750226312091001
Mean40:  99.99784699275877
StandardError40:  2.3723878186728786
```

The heights of the bars represent the relative frequency of the various sample means for n = 10 and n = 40. The means and standard errors for both distributions are incredibly close to the theoretical values. Both sampling distributions center on the population mean of 100. However, the n = 40 distribution is noticeably tighter than broader one for n = 10. These characteristics match our theoretical expectations.

What are the practical implications of this difference?

The tighter sampling distribution indicates that the sample means cluster closer to the actual population mean for the larger sample size. Notice how the wider n = 10 spread has more sample means farther away from the population mean (100). If you use a sample size of 10, you're more likely to obtain sample means that are more erroneous than with a sample size of 40.

For instance, by looking at the bars' heights, obtaining a sample mean of 92 or even further away from 100 for n = 10 is entirely possible. In comparison, obtaining a sample mean that far off for n = 40 is very unlikely because you can barely even see the bars at those values.

Sample sizes are not just a statistical nicety but a fundamental component of trustworthy research. Variability is an inherent population characteristic that the researchers typically can't reduce to obtain more precise estimates. Hence, larger samples are the practical method for increasing precision. As you increase the sample size, the difference between your sample mean and the population mean tends to decrease. In other words, larger sample sizes produce more precise estimates!

I'm sure you already knew this old statistical adage, but now you see why that's the case. While this section focuses on sample means, you can apply the same ideas to differences between groups, proportions, and regression coefficients.

Finally, the predicted values for the mean and standard errors in the simulation virtually match the values we calculated using statistical theory. Hence, statisticians can estimate a sampling distribution using data from a single sample. What good is that?

Recall that in Chapter 3, I said that sampling bias was worse than random sampling error because statistical procedures can estimate and account for the random error but not the bias. If our apple samples are biased (e.g., the apple scale reads too high), we couldn't determine that from the data alone. We'd need outside information, such as a calibration study using known weights.

However, in the apple weight example, we see that a single *unbiased* sample estimates the sample mean's precision because the theoretical values using the formula match the simulation results.

In short, statistical methods can use a single sample to estimate how close a sample statistic is to the actual population value. The following section covers several tools that do just that. Fortunately, researchers don't need to collect many random samples.

Confidence Intervals and Margins of Error

While estimates are always wrong to some extent, they are more informative when they're likely to be close to the correct value. Very wrong estimates are not helpful.

We never know the correct population value exactly, but we need a way to quantify how wrong sample estimates are likely to be.

Confidence intervals (CIs) are handy tools for estimating the range within which a parameter is likely to fall. They use sampling distribution information to produce a range of likely values for parameters, such as the population mean and proportion.

CIs provide a margin of error around a sample estimate, so we have an idea of how wrong it is. Similarly, the margin of error in a survey tells you how near you can expect the survey proportion (e.g., proportion voting for a particular candidate) to be to the correct population value.

For example, imagine a new medication supposedly increases people's IQ. One study found that the medicine produced an average increase of 7 IQ points in the sample. That sounds good. But its confidence interval is [1 13]. What does that mean?

If the whole population took the medication, the average increase would likely be between 1 and 13 IQ points. That range incorporates the uncertainty due to random sampling error, variability, and sample size. That range goes from a trivial increase of 1 IQ point to a substantial increase of 13 points. The study suggests an effect exists, but there's enough uncertainty that we can't determine whether it's a small or large effect. The considerable uncertainty limits the usefulness of the results.

Now, suppose a later study assesses the same medication in the same population but uses a larger sample size. This study also finds an average increase of 7, but its confidence interval is [6 8]. We can be confident that the population mean falls within that range. It's a much smaller range, indicating a more precise estimate. There's less uncertainty. The results of this study provide a much clearer picture of the drug's effectiveness.

Both studies found the same mean increase of 7 IQ points, but the results of the second study were much more informative. This example further shows how the mean by itself provides only partial information. The confidence intervals factor in the random variability to indicate how close the sample effect is likely to be to the population effect.

The margin of error you see in polls plays a similar role. A survey will find a particular proportion of respondents who prefer a specific candidate or favor a policy. The margin of error indicates how close the sample proportion is likely to be to the population proportion.

For both confidence intervals and margin of error, smaller ranges indicate the sample estimates are more precise and are likely to be close to the correct population value.

Finally, remember the caveat about these tools and bias. Typically, statistical procedures can handle random error but cannot detect bias. This limitation applies to both confidence intervals and margins of error.

Statistical Significance

The previous CI example relates to the concept of statistical significance. While analysts typically determine significance using p-values, it's easier to understand using CIs.

For starters, what does statistically significant mean?

Let's revisit the hypothetical IQ drug example. In this scenario, the value of zero indicates that the drug has no effect because it represents an increase of zero IQ points. That's the null hypothesis. Frequently, zero represents the null in a test.

Suppose a study tests this medication and finds an average increase of 4 IQ points. The CI is [-4 12]. While the sample estimate is 4, the range of likely population values extends from -4 to 12. Critically, this interval includes zero or no effect. While there could be an effect, the estimate isn't precise enough to know for sure because the CI also contains no effect. In statistical lingo, the results are not statistically significant. You fail to reject the null hypothesis.

In this scenario, the study's p-value will be greater than the significance level (e.g., 0.05). There's too much variability, and our estimate is too imprecise to conclude that the medicine will increase IQ points in the population even though the sample mean is +4 points. That sample mean might be due to random sampling error alone.

Suppose another IQ study finds a mean effect of 7 IQ points and a CI of [6 8]. This confidence interval does not include zero. The entire range of likely values suggests an effect exists in the population. These results are statistically significant, and you can reject the null.

The p-value for this study will be less than or equal to the significance level. While there is still some uncertainty about the average difference at the population level, the entire range of values agrees that the medication increases IQs.

The relationships between p-values and confidence intervals are the following:

- When the CI includes the null (no effect) value, the p-value will be greater than the significance level. You fail to reject the null hypothesis. The results are NOT statistically significant.
- When the CI excludes the null, the p-value is less than or equal to the significance level. You reject the null, and your results are statistically significant.

In short, statistical significance indicates that an effect you see in sample data likely also exists in the population after accounting for random sampling error, variability, and sample size. Your results are statistically significant when the p-value is less than your significant level or, equivalently, when your confidence interval excludes the null hypothesis value.

Conversely, non-significant results indicate that despite a sample effect, you can't be sure it exists in the population. It could just be random error.

I prefer confidence intervals because they indicate both statistical significance and the estimate's precision. You can use them together. Confidence intervals and p-values will always agree when you use the appropriate methods.

Benefits of a Large Sample Size

Let's recap the benefits you gain by having a larger sample size.

Increased Precision

The sample size affects the width of sampling distributions. Larger samples tend to produce narrower CIs, indicating more precise estimates of the population parameters.

Estimate precision refers to how closely sample statistics align with the population parameters. A larger sample size tends to yield more precise estimates because it reduces the effect of random fluctuations within the sample. With larger samples, the positive and negative values more effectively cancel each other out, narrowing the margin of error around the estimated values.

Narrower curves equate to greater precision. Sample estimates will have narrower confidence intervals, smaller margins of error, and tend to be close to the correct population value.

For example, estimating the average height of adults using a larger sample tends to give an estimate closer to the actual average than using a smaller sample.

Greater Statistical Power

Statistical power is the probability that a study will detect an effect when one exists, such as a difference between groups or a correlation between variables. Larger samples increase the likelihood of detecting actual effects.

In other words, if you use a smaller sample, it is more likely to miss an effect that exists in the population.

Better Generalizability

With a larger sample, there is a higher chance that the sample adequately represents the diversity of the population, improving the generalizability of the findings to the population.

Consider a national survey gauging public opinion on a policy. A larger sample captures a broader range of demographic groups and opinions.

Reduced Impact of Outliers

In a large sample, outliers have less impact on the overall results because many observations dilute their influence. The numerous data points stabilize the averages and other statistical estimates, making them more representative of the general population.

If measuring income levels within a region, a few extremely high incomes will distort the average less in a larger sample than in a smaller one.

Limits of Larger Sample Sizes

While larger sample sizes offer numerous advantages, it's essential to understand their limitations. As you learned in previous chapters, large sample sizes are not a panacea for all research challenges. Ignoring these issues can lead to misleading conclusions, regardless of how many data points are collected.

Sampling Bias

Even a large sample is misleading if it's not representative of the population. For instance, if a study on employee satisfaction only includes responses from headquarters staff but not remote workers, increasing the number of respondents won't address the inherent bias in missing a significant segment of the workforce.

Other Forms of Bias

Biases related to data collection methods, survey question phrasing, confounders, or data analyst subjectivity can still skew results. If the data collectors don't address those issues, a larger sample size might magnify these biases instead of mitigating them.

Errors in Study Design

Simply adding more data points will not overcome a flawed experimental design. For example, increasing the sample size will not clarify the causal relationships if the design doesn't control a confounding variable.

Large Sample Sizes are Expensive!

Additionally, it is possible to have a sample size that is too large. Larger sizes come with challenges, such as higher costs and logistical complexities. You reach a point of diminishing returns where a huge sample will detect such trivial effects that they're meaningless in a practical sense.

The takeaway is that researchers cannot rely solely on large samples to safeguard the precision and validity of their results. They must use an appropriate sampling method, a sufficiently large sample size, a robust study design, and meticulous execution to understand and accurately represent the phenomena.

Examples: Small Sample Size Problems

In statistical analysis, small sample sizes present numerous challenges that can significantly compromise the validity of research findings. Small samples exhibit the opposite characteristics of large samples— low statistical power and less precise estimates. In other words, they're more likely to miss existing effects, and their sample estimates (e.g., the mean) tend to be further from the correct population values.

Additionally, conducting reliable subgroup analyses is problematic due to the need for more data within each subgroup. This shortfall can yield misleading conclusions about a variable's impact on different subgroups, as small subsets might inadequately capture the diversity of the population.

Let's focus on the increased variability of results from small samples, a critical but often overlooked issue. Think back to the wider sampling distributions associated with small samples. Those curves extend far from the mean in both directions indicating they're more likely to produce unusually high *and* low means.

Small samples can lead to high variability in research results, which means that the findings from one study might differ substantially from another, even if they are exploring the same phenomenon. This variability often misleads researchers and stakeholders into drawing overly confident conclusions from inadequate data.

One of the psychological biases contributing to this issue is the tendency to believe that small samples closely resemble the entire population they represent. This misconception is part of a broader human tendency to overestimate the consistency and coherence of our observations.

In research, this can manifest as an exaggerated faith in the conclusions drawn from just a few observations. This bias is closely related to the halo effect, an additional cognitive bias to those that I covered in the first chapter. The halo effect is where we assume we understand and know more about a subject or person based on minimal information. This mental error pushes us to give too much weight to conclusions drawn from a small sample size.

Let's look at several real-world examples where small samples led researchers astray.

The Small Schools Movement

The small schools movement in education provides a striking example of how small sample sizes can mislead policy decisions. Observations that some small schools performed exceptionally well led to significant investments in reducing school sizes. The rationale was that smaller schools fostered better educational outcomes.

However, it was later observed that the performance variability among small schools was high—some showed markedly worse performance than their larger counterparts. This phenomenon was an artifact of extreme outcomes in both the positive and negative directions prevalent in small samples thanks to their wider sampling distributions.

Kidney Cancer Rates in Rural Counties

A similar illustration of the problems associated with small sample sizes can be seen in the analysis of kidney cancer rates in rural counties, as discussed in Daniel Kahneman's book "*Thinking, Fast and Slow.*" Initially, researchers observed that the counties with the lowest rates of kidney cancer were rural and sparsely populated, leading to the premature conclusion that rural lifestyles might promote health.

Paradoxically, the counties with the highest incidence of kidney cancer were also rural and sparsely populated. This contradiction was explained away by hypothesizing poor access to healthcare and unhealthy lifestyles as potential causes. However, both extremes were artifacts of the small sample sizes involved in rural areas. These examples underscore that small sample sizes are more likely to exhibit extreme outcomes (high and low) simply due to the high variability inherent in small samples.

Replicability

One significant challenge posed by small sample sizes is the issue of replicability. For instance, the small schools movement initially suggested that smaller schools were inherently superior, based on a few high-performing examples. However, when researchers tested these findings more broadly, the results were not replicable, revealing that performance varied greatly among small schools. Similarly, the analysis of kidney cancer rates in rural counties presented conflicting results that were not replicable across similar demographic settings.

These anomalies were primarily due to small sample sizes, which were more susceptible to extreme values. The inability to replicate findings consistently is a critical concern, as replicability is a cornerstone of scientific validity. Studies based on small samples often produce unique or significant-seeming results that, upon broader examination, prove to be artifacts of the limitations inherent in small sample sizes.

Indeed, the abundance of small sample sizes in particular fields, such as psychology, partially explains the replicability crisis I discussed in Chapter 6. Small samples raise questions about the validity of specific studies and can erode confidence in entire research areas.

Sample Size Summary

Samples should be large enough to adequately represent the population and have a reasonable chance of detecting a meaningful effect. The sample size is a researcher's primary method for managing the population's variability.

Recognizing the limitations and challenges of small sample sizes is crucial for researchers to avoid these common traps. While smaller studies can and do provide valuable insights, analytical thinkers must interpret their findings cautiously, particularly when making broad generalizations or policy decisions. In statistical research, an awareness of the increased variability induced by small samples can lead to more robust and reliable science.

Additionally, recall the Bizarreness Effect—that cognitive bias that causes us to recall distinct (i.e., bizarre) results more strongly. New, unusual findings are more likely to be reported, discussed, and remembered. Small samples are more likely to produce that type of result, overly weighting them in our consciousness.

Understanding the role of sample size is crucial in both experimental and big data contexts. Larger samples provide more reliable and precise estimates, while smaller samples can compromise validity.

For researchers conducting experiments and those evaluating such studies, it's essential to know how sample size influences the results' precision and generalizability. While the large datasets associated with Big Data mitigate some concerns, they are not immune to systematic biases.

Thus, whether a dataset is large or small, an awareness of these dynamics is essential for robust analysis. This understanding helps in not only conducting sound research but also in critically reviewing studies for informed decision-making. Always remember that the breadth of the sample profoundly influences the conclusions' strength.

As we continue through the rest of this chapter, we'll explore more ways variability makes data analysis more complex.

Regression to the Mean

Regression to the mean is an odd effect related to variability that can trip up analytical thinkers when making real-world observations and conducting more formal analyses.

It is the statistical tendency for an extreme sample or observed value to be followed by a more average one. It is the propensity for a later observation to move closer to the mean after an extreme value. The concept applies only to random variation in a process or system and does not pertain to interventions or events that affect the outcome.

In short, outliers and flukes are likely to be followed by more typical events.

For example, suppose classes of students take a standardized test, and the average score for a class is 75%. If the students in one class average 90%, the next class will likely be lower and closer to the mean. Conversely, if a class averages 60%, the next class will likely be higher.

Regression to the mean is an important concept that explains why more average results often follow extreme ones in subsequent measurements. It's a natural occurrence requiring no intervention. If you observe something unusually high or low, chances are it will be closer to the average the next time you see it.

It is not just a theoretical concept but a practical one with significant implications. Whether it's a surprising performance in sports, a sudden market shift in stocks, or a change in standardized test scores, the concept of regression to the mean is at play, bringing results closer to the average.

Let's learn more about this intriguing statistical concept and why it occurs, explore its implications, and why it matters in research.

Examples

The following examples illustrate regression to the mean in various contexts. In all these cases, someone might search for an underlying cause that produced the change, but it could just be a natural byproduct of random variation. Consider this phenomenon in your efforts to think analytically in the everyday world.

Sports Performance

Sports can be an excellent source for regression to mean with all the statistics they record!

A basketball player scores unusually high in one game but returns to their average scoring in the subsequent match.

The idea can also apply to entire seasons because a season is a sample of games. A rookie can have an outstanding season and then experience the "sophomore slump," where performance declines toward the average. Conversely, baseball players with subpar batting averages one season tend to improve toward the mean the following season.

The "Sports Illustrated Cover Jinx" is often cited as an example of regression to the mean. Athletes who perform exceptionally well to merit a cover story typically see their performance decline afterward, not due to a curse but simply returning to their usual performance level.

Investment in High-Performing Stocks

An investor selects a range of stocks for their portfolio because they performed unusually well in one quarter. However, in the following quarter, their performance regresses to more average levels, reflecting typical market behavior. Regression to the mean in action!

Hospital Emergency Room Visits

Over a week, a hospital's emergency room sees significantly fewer patients than average, seemingly without any specific cause. The following week, the number of patients returns to the usual average, aligning with the typical random variations in ER visits.

Standardized Test Score Policy

Massachusetts's 1999 effort to improve standardized test scores is an example of regression to the mean. That year, schools were given goals to improve their average test scores. Many of the lowest-performing schools achieved their targets. The policy looked successful at first. However, many top schools failed to meet their goals. That's regression to the mean, where extreme scores in both tails of the curve naturally move towards to the central average.

Praise vs. Criticism

From fighter pilots to educational settings, trainers often criticize poor performers and praise high performers. Observers have noticed that the worst performers do better subsequently while the top ones do worse. This pattern leads to the mistaken conclusion that criticism boosts performance more than praise.

By now, you can see that this pattern was likely due to regression to the mean!

Research Fallacy

The regression to the mean fallacy arises in research when analysts mistakenly attribute changes in the outcome to manipulations of an experimental factor rather than observations reverting to the mean. This fallacy overlooks how unusual outcomes shift towards the mean due to random chance, producing an apparent improvement.

To illustrate, consider a study in which researchers measure blood pressure and select individuals with unusually high readings. That's the group they want to focus on. The initial higher-than-average readings could partly be due to random chance, making the underlying condition appear worse. Consequently, when they retest these individuals later, their blood pressure readings are likely to shift towards more typical levels, creating the appearance of an improvement.

If researchers hastily conclude that this change is entirely due to their intervention, they fall prey to the regression to the mean fallacy. The subsequent reduction could be partially due to natural, random fluctuations in blood pressure. That random chance is separate from any intentional changes in the study's variables.

Experiments targeting subjects with extreme characteristics must include a control group with similar extreme traits to account for this phenomenon. This approach helps differentiate between effects due to the intervention and those resulting from regression to the mean.

Additionally, repeated measurements before the experiment can help determine whether the initial extreme values were random fluctuations or part of a consistent trend. That's better than relying on a single extreme measurement.

Understanding and accounting for the regression to the mean fallacy is a crucial aspect of research. It's about avoiding the trap of drawing incorrect conclusions from data simply regressing to its mean.

Worked Probability Example

You've seen the regression to the mean examples and know it occurs due to random variation, but how exactly? Let's take a look.

First and foremost, this propensity exists due to probabilities relating to sampling distributions. It does not reflect a memory or intentional adjustment in the system. People can get confused because it seems to counter the Gambler's Fallacy by invoking a balancing mechanism that remembers past events. But it doesn't.

Regression to the mean results from a consistent distribution of values where extreme values are less likely to occur than more central values. Consequently, probability distributions can model this phenomenon.

Let's see how regression to the mean occurs with a worked example. I'll use the distribution of IQ scores, which follow a normal distribution with a mean of 100 and a standard deviation of 15.

Imagine that researchers draw 5 participants randomly from this distribution and calculate a mean IQ of 90. We can use a sampling distribution of the means to calculate the probability that the following sample of 5 will have a mean closer to the population mean of 100.

The following sampling distribution of the mean is for IQ scores with a sample size of five. Because it is a sampling distribution, each point on the curve relates to sample means, not individual values. I'm performing a two-tailed analysis because a sample mean of 110 is equally extreme as a sample mean of 90, just in the other direction from the mean.

The shaded region in the graph ranges from 90 – 110 and represents less extreme sample means than our sample mean of 90. Visually, you can see how the bulk of sample means will be less extreme because the shaded area is much larger than the two more extreme regions. That, in a nutshell, is how regression to the mean works.

But I also love how these probability distributions can quantify it! This graph shows that the probability of the second sample being less extreme is 0.8639.

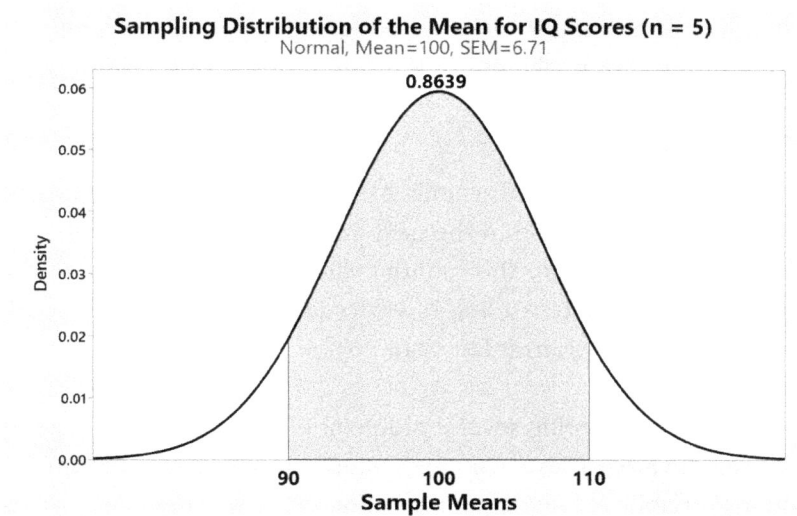

Sampling Distribution of the Mean for IQ Scores (n = 5)
Normal, Mean=100, SEM=6.71

Comparing the probability of the second sample being less extreme (0.8639) to it being at least as extreme (1 – 0.8639 = 0.1361) reveals that it is 6.3 times more likely to be less extreme, as shown in the following formula.

$$\frac{0.8639}{0.1361} = 6.3$$

A sample mean of 90 might not seem extreme compared to a population mean of 100, but you can already see regression to the mean in action.

Now, suppose the first sample mean is 85, a bit more extreme. Using the same calculation process, the following sample mean is a whopping 38.4 times more likely to be less extreme!

Regression to the mean is real even though it relies on random chance. It's the natural outcome for distributions where extreme values are less likely than central values. The worked example shows the tendency can be substantial. Regressing to the mean is powerful enough that you must account for it when designing experiments using subjects with extreme attributes.

Variability Over Time

Data often comes at you over time. And it seems like a simple task to compare one observation to the next and draw conclusions. For example, you can compare this month's sales to the previous month's, today's blood pressure reading to yesterday's, or this quarter's GDP growth to the same quarter last year.

Hopefully, you're seeing several problems with the above process. It has a sample size of just two data points, and you must account for random variability. You need more information to draw conclusions about business growth, health, or economic change. Two points don't make a trend!

Your blood pressure is lower today than yesterday, but is that change random variation or a trend? With just those two observations, it's impossible to tell.

In all these cases, you need a larger context. A good solution is to track an outcome over time so you can monitor its fluctuations. Gathering more data points helps you see a trend more clearly and better understand the inherent process variability. Understanding the expected variability in a process over time enables you to recognize changes more quickly. Sometimes, you can identify a change with a single new data point when it falls outside the normal process variability. This quick detection allows the process owner to respond more quickly to problems.

Conversely, by understanding the normal process variation, the process owner won't overreact to normal fluctuations that are entirely random. For these reasons, the quality improvement field uses control charts for manufacturing and business processes. However, I'm a big proponent of using them in other contexts, such as research.

Imagine we're conducting a height study. Our study's random sampling procedure and measurement methodology are also processes. As our project records heights, we record them and graph them in a control chart. If we suddenly start to see warnings in these charts, we know something is wrong.

Either something has gotten into the water supply and started affecting heights, or, more likely, our random sampling procedure has gone awry, or a problem occurs during the measurement process. We'd need to investigate to find the source of the problem. But the control chart tells us more quickly that something in our process isn't working, and we can make adjustments rapidly. That can save us time and money.

Let's review control charts in their traditional process control setting, and then I'll switch gears and show you how they can be beneficial in other contexts.

Control Charts

Control charts determine whether a process is stable and predictable or out of control and needs adjustment. Some degree of variation is inevitable in any process. Control charts, or Shewhart charts, help prevent overreactions to normal process variability while prompting quick responses to unusual variation.

They plot process data and help you identify common cause and special cause variation. These graphs can determine whether the process is stable and if variability is a problem. Control charts can determine whether variability is intrinsic to the process or related to specific sources (i.e., special causes). By identifying the different sources of variation, you can keep your process stable without over-correction.

A stable process operates within an ordinary, expected range of variation. It is predictable and consistent, and special causes of variation don't influence it. Special causes include the following:

- Changes in the process itself.
- Changes in the environment.
- Changes in the input materials or equipment.

Stable processes are more likely to produce high-quality products or services. Conversely, an out-of-control process is unpredictable and more likely to cause defects or errors.

A control chart displays process data by time and upper and lower control limits that delineate the expected range of variation for the process. These limits let you know when unusual variability occurs. Statistical formulas use historical records or sample data to calculate the control limits. Unusual patterns and out-of-control points on a control chart suggest that special cause variation exists.

Control charts can be valuable aids for tracking a continuous process and gaining insight into a newly established one. They can help with the following:

- Determine whether a process is stable.

- Find problems as they occur in an ongoing process.
- Assess the effectiveness of a process change.
- Predict the range of outcomes for a process.
- Assess patterns of special cause variation to identify non-routine events.
- Determine whether improvements should target non-routine events or the underlying process itself.

For example, quality engineers at a manufacturing plant monitor part lengths. They use process data to create an X-bar-R chart, a control chart that evaluates both the process mean (X-bar) and spread (R chart for range).

Xbar-R Chart of Length

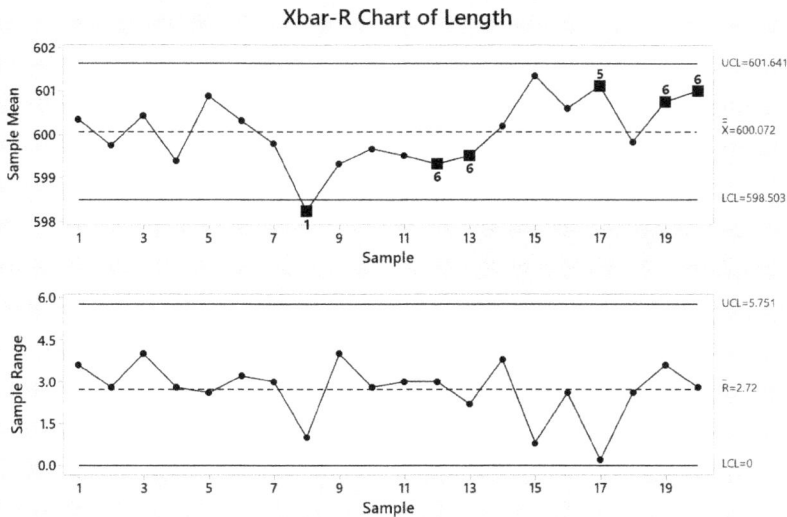

Control charts typically contain the following elements:

- Data points representing process outcomes.
- Control limits depict the range of normal process variability.
- Centerline locates the process's center value.
- Bolded squares with numbers indicate out-of-control points.

Interpretation

For the part length example, we must ensure the R chart (bottom) is in control before analyzing the X-bar chart. If the R chart is unstable, the control limits for the X-bar chart will be invalid, potentially leading to false signals of an out-of-control situation on the X-bar chart.

The R chart does not flag any points. They're all in control. However, the X-bar chart on the top is a different story because it flags six points. Bolded, square data points with numbers failed a statistical test, suggesting that special cause variation exists.

Point 8 is out of control because it is below the lower control limit. But there are five more out-of-control points within the control limits. Why?

Control charts can test for various statistically improbable patterns, some of which are subtle enough that we might not notice on our own.

The chart flags points 12, 13, 19, and 20 because 4 out of 5 points in a row are more than one standard deviation from the centerline on one side of the mean. That's unlikely to occur by chance. Additionally, #17 is flagged because 2 out of 3 points are more than two standard deviations from the centerline on one side of the mean.

All out-of-control dots suggest special cause variation exists because those patterns are unlikely to occur with only common cause variation. Assessing these patterns with process knowledge will help us identify their source.

These results tell the process owner they should find the sources and make corrections.

When these graphs determine that a process is stable, you can perform additional analyses to draw conclusions about the process. However, an unstable process is unpredictable, and you can't draw reliable conclusions about its behavior. Any conclusions that you draw today might not be correct tomorrow.

Control charts are linked to business processes, but these plots provide tremendous benefits for processes and hypothesis testing outside the quality improvement realm. Spoilers—control charts check an assumption we often forget about for hypothesis tests.

Control Charts can Assess Non-Business Processes

The trick to seeing how control charts work in a wide variety of settings is to enlarge your notion of processes to include non-business processes. After all, instability and variability are problems in many other environments. For instance:

- Teaching is the process of transferring knowledge that is measured by testing.
- People with diabetes have a process for maintaining blood sugar at a stable level.
- I had a process for causing research participants to experience impacts of 6 times their body weight.

These processes can be unstable or stable, have some natural variability, and might have special causes of variability. Assessing these issues can help you improve the processes. Just like business processes, if your data aren't stable, the conclusions you draw using hypothesis tests are unreliable.

Control charts verify the assumption that a process is stable. We don't usually think of applying this assumption to hypothesis tests. However, data for a hypothesis test must also be stable. Otherwise, the conclusions aren't reliable.

Imagine you're comparing the means of two groups. If one of the groups is unstable, any conclusions you draw are meaningless because they'll vary based on timing. You'll get different results if you conduct the same study slightly later because the group is unstable and changing.

Let's return to my bone density study and I'll show you how control charts gave crucial information about this non-business process.

Example: Using a Control Chart in My Study

Earlier in this chapter, I discussed a bone density study where our study had middle school participants jumping off 24-inch steps 30 times on alternating school days. The research goal was to determine whether these impacts would cause their bone density to increase. We defined the treatment as impacts of six times their body weight.

Unfortunately, while the mean of the impacts was above six times the body weight, almost half the subjects did not experience impacts of this magnitude. We knew this was not good enough. All subjects should achieve the target impact force.

To devise a solution, I conducted a pilot study and plotted the data from this process on an Xbar-S control chart.

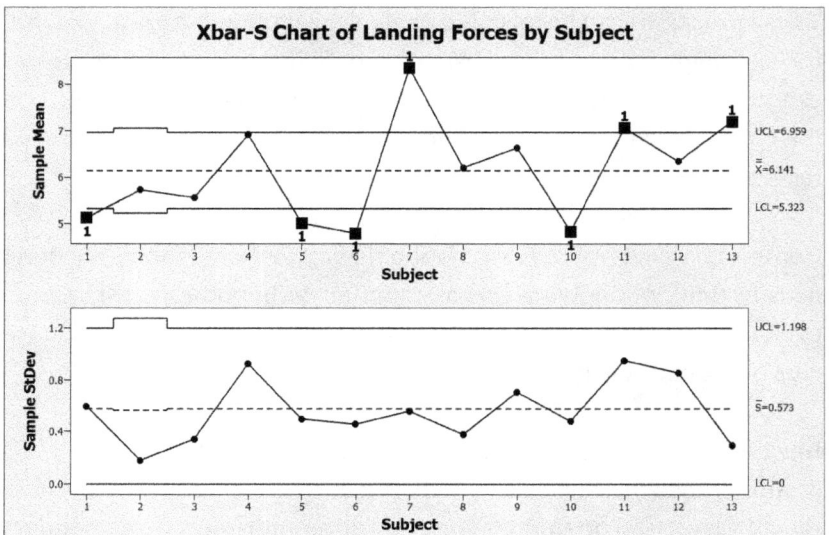

To interpret this chart, start by looking at the S chart at the bottom, which displays the variability of each subject's landing impacts. No points are outside the control limits, and the subjects' variability is similar. This graph indicates that each subject has a consistent, predictable landing style, which helps us rule out natural variability within each participant as the problem.

The Xbar chart on the top shows that the overall mean (6.141) is greater than our target. Unfortunately, data points fall outside the control limits, indicating this process is unpredictable. Different subjects have dramatically different average landing impacts.

Taken together, the control chart indicates that some participants consistently have large impacts while others consistently have lower ones. However, this variability is not intrinsic to the process (common cause variation) but assignable to differences between the participants (special cause variation).

This information guided the corrective measures that we implemented. Had the variability been inherent in the process, we probably would have built higher steps to increase the mean. However, because we could attribute the variability to the subjects, we decided to teach the subjects how to land and have a nurse watch all jumping sessions to provide feedback on the spot. This combination lessened the variability between subjects enough so that all impacts were greater than six body weights.

Success! Even though this was not a business process, a control chart provided invaluable information.

Adjusting a Mean is Easier than Reducing Variation

The jumping example illustrates a generally understood concept in quality improvement—it is often easier to change the mean than to reduce the variation. But the idea applies to other contexts.

The original jumps ranged from 4.7 to 8.4 BWs. That's a wide range, representing high variability between subjects. Some subjects received nearly twice the "dose" as others. If we had raised the steps, we could've easily increased the mean impacts to be further above 6 BWs. However, the data suggest they'd still have individual landing styles that would've produced significant variability. By teaching a standard landing style that eliminated the different styles, all the subjects were above 6 BWs, and their impacts were closer together (lower variability). That's a more consistent dose.

However, teaching the standard landing style and then observing and reinforcing it during the intervention sessions involved much more work than simply building a higher jumping platform.

Experts in quality management acknowledge that managing variability typically presents a greater challenge than adjusting the mean. To alter the mean, changing a manufacturing setting or target is sufficient. Conversely, reducing variability usually demands adopting improved technologies or new processes.

For example, if you want your oven to cook at a higher mean temperature, you can turn its thermostat up. However, to reduce the temperature variability during cooking, you'll need a higher-quality oven.

Similarly, if you want more precise measurements (low variability around the correct value), you almost always need to use more expensive measuring instruments.

This issue is yet another way variability is problematic. Reducing variability is usually more challenging than changing the mean.

Closing Thoughts

Random Variation is a Form of Uncertainty

Let's take a moment to reflect on what we've covered because I want to emphasize the fundamentality of variability. My goal here is to provide a thorough understanding of why fully understanding the role of variability is critical for all analytical thinkers.

We've delved into the concept of variability and the many ways it impacts our lives and data analysis. Variability is not just a number but a fundamental aspect affecting the precision and reliability of statistical conclusions. Various examples show how variability can complicate analyses.

One of the critical points is the distinction between central tendency and variability. While measures like the mean, median, and mode provide a snapshot of data's central value, it is the variability or the spread of data that often has consequences. Recall the example of two pizza restaurants with the same average delivery time but different levels of variability in delivery times, illustrating how variability can drastically affect service reliability and customer satisfaction.

The chapter also introduced more complex statistical concepts like sampling distributions and the role of sample size in reducing the standard error, thus enhancing the precision of sample estimates. Understanding these distributions is crucial for knowing how sample size affects your results. Increasing the sample size directly reduces the variability of sample estimates, producing sample estimates that are closer to the correct population values.

Large sample sizes produce more precise estimates. They can cut through the noise of random variability. Small samples are more susceptible to noisy data and are prone to delivering extreme results that are not reproducible.

Further, understanding the implications of variability when drawing conclusions is crucial. Variability influences the ability to detect genuine effects in the data, likening it to the signal-to-noise ratio in statistical analysis. High variability, or noise, can make it difficult to discern the actual signal, necessitating larger sample sizes. It can also lead to misleading conclusions, suggesting extreme effects where none exist or obscuring real effects.

Fortunately, confidence intervals, margins of error, and p-values can help us separate true population effects from random sampling error. Just because we see an effect in our sample doesn't mean it exists in the population. Confidence intervals help us understand the precision of our sample estimates.

In short, I want you to view variability as uncertainty. If there were no variability, you'd be absolutely sure about the pizza delivery time, your drive time, the effectiveness of your medication, and so on. As variability increases, all these outcomes become less certain, and it'll be easier to draw incorrect conclusions. Statistical analyses will have a harder time identifying significant differences.

Then, we considered the intriguing concept of regression to the mean and how it can trip up conclusions involving unusual subpopulations in everything from everyday observations to scientific experiments.

Finally, we looked at variability over time and how understanding it using control charts can help you respond to real process changes more quickly while avoiding the trap of overcontrolling a process by reacting to inherent random variability.

In conclusion, understanding and accounting for variability is vital in data analysis. This chapter underscores the importance of considering variability alongside central tendency and equips you with the knowledge to anticipate and mitigate its effects in practice.

By understanding both the mean and variability, analysts can derive more meaningful insights from their data, ensuring that their conclusions genuinely reflect the underlying population characteristics. This holistic approach to data analysis is essential for making informed decisions.

The Hardship Fallacy in Statistical Significance

The Hardship Fallacy is a term I've coined to describe an attitude toward statistically significant results that researchers find in challenging research environments. It combines several concepts you've read about thus far in the book.

Researching and analyzing data to find significant results can be difficult. Often, even great ideas don't show clear outcomes, and achieving statistical significance is tricky. This challenge is amplified in fields like education and psychology, where the study subjects are inherently varied and difficult to measure consistently. Additionally, collecting enough data can be expensive and time-consuming in these settings. These conditions frequently cause studies in these areas to have small sample sizes with high variability.

Sometimes, researchers succeed in finding statistically significant results under these demanding conditions. It can be tempting to think that if a finding emerges as significant despite high noise and small samples, it must reflect a powerful underlying effect. However, this reasoning is flawed. In reality, statistically significant outcomes derived from such noisy circumstances are less reliable. They often overestimate the actual effect or even wrongly predict its direction.

A common misconception is that difficult conditions make a significant result even more reliable. This misunderstanding overlooks the fact that it's harder to determine the actual effect accurately in noisy studies with small samples. This error in reasoning can mislead even seasoned researchers, leading them to trust results that they should view with extreme caution. It's essential to understand that impressive findings under challenging conditions do not necessarily suggest more potent effects but instead show the problems of noisy data and small samples.

Studies have a low capacity for producing correct results when dealing with these conditions. Any statistically significant estimates these studies find must inherently appear "substantial." For these estimates to stand out from the noise, they require a considerable effect size. However, that appearance is more likely to occur in these noisy conditions due to chance than a genuinely substantial effect.

A thinking analytically mindset requires you to recognize when to be extra skeptical about analytical conclusions. Don't fall victim to the Hardship Fallacy. Be aware of results obtained under noisy data conditions!

Read my related article on this topic:
https://statisticsbyjim.com/hypothesis-testing/low-power-studies/

CHAPTER 8

Analytics: Multivariate Complexities

For the final chapter, we'll stay on the Analytics level of the Thinking Analytically pyramid.

Most problems in the real world involve multiple variables. That probably doesn't surprise you. However, it introduces complexities in how you analyze data and interpret the results. You've seen some of that in the discussions about managing confounding variables and the bias problems that occur when researchers don't control them.

In this chapter, we'll focus on several of these multivariate complications. I won't get deeply involved in how to fit and evaluate regression models. There are entire books about that! But we'll look at various types of relationships analysts need to model. As with confounding variables, your results can be untrustworthy if you don't model them correctly.

You'll also see how regression and other forms of modeling can be seductive! Yes, you read that right. Analysts will face temptations that are hard to resist, even though they can produce bad outcomes.

I bet you never thought there'd be a bit of melodrama in a book about thinking analytically! However, the number of options and the pursuit of better-fitting models allow subjectivity and the various cognitive biases from Chapter 1 to creep in. Something to be aware of!

At the end of the chapter, we'll look at how the recent advances in AI fit in with the Thinking Analytically pyramid.

Confounding Variables

We covered confounding variables in depth earlier on while talking about experimental designs. Using a randomized design is a huge benefit because you don't need to worry about modeling known and unknown confounders. You can essentially compare the treatment and control means directly and see if they're significant. That's simplifying it because you can use more complex models. But you don't need to worry about measuring and controlling confounders, which is a huge deal.

However, confounders are a genuine concern for observational data.

When confounders are a concern, the ideal case is that simply including a confounder in the model should remove its bias. In Chapter 5, you saw how the subject's weight confounded the relationship between activity and bone density. I just had to include weight in the model to reveal the proper relationship between activity and bone density. Simple.

Unfortunately, the possible presence of confounders also opens the door to potential problems. Suddenly, you need to measure and include known confounders in your model. And what about unknown confounders?

Additionally, it gives analysts an excuse to tinker with the variables they include in the model, trying to find the best fit. Think back to the p-hacking discussion where fitting an excessive number of models is a bad practice that can increase the potential for falsely positive significant results.

Model tinkering can morph into attempts to improve how well the model fits the data. From the analyst's perspective, best practices require them to investigate and try variables that don't relate to their study other than the fact that they're potential confounders. Leaving them out of the model would bias the results, while including them removes the bias.

Consequently, exploring potential confounders sounds like a great thing, but taken to extremes, it tends to inflate false positives. Again, imagine all the cognitive biases that come into play surrounding which variables to include when they happen to support the analyst's preexisting hypothesis or ability to get a publishable result.

The way to avoid this problem is by letting theory guide your choices.

- Which variables are logical confounders?
- If including a variable improves the model fit, does theory suggest it makes sense?
- Do the overall results of the model fit theory?

There will be more on this later in the chapter, but keep this idea about model tinkering in mind as we discuss additional types of effects that analysts can model.

Curvature

In regression analysis, curve fitting is the process of specifying the model that provides the best fit to the specific curves in your dataset. Curved relationships between variables are more complex to fit and interpret than linear relationships.

For linear relationships, as you increase the explanatory variable by one unit, the mean outcome always changes by a specific amount. This consistency holds regardless of where you are in the observation space.

Unfortunately, the real world isn't always nice and neat like this. Sometimes, your data have curved relationships between variables. In a curved relationship, the change in the outcome associated with a one-unit shift in the explanatory variable varies based on the location in the observation space. In other words, the effect or relationship is not a constant value.

The following fitted line plot is a simplified example of using a linear relationship to fit a curved relationship. The line represents the model and it doesn't fit the curvature in the data.

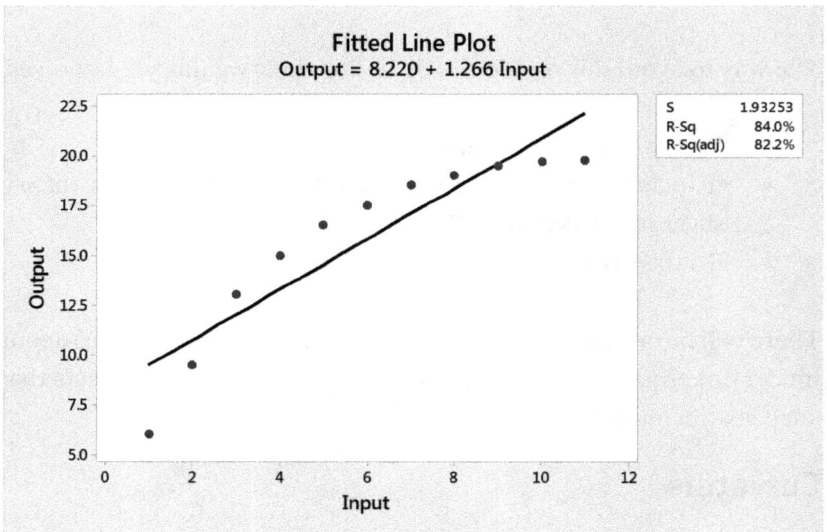

Fitted Line Plot
Output = 8.220 + 1.266 Input

S	1.93253
R-Sq	84.0%
R-Sq(adj)	82.2%

R-squared is the standard goodness-of-fit measure for many regression models. Higher percentages indicate the data fit the model better. However, as you'll see throughout this chapter, chasing a high R-squared creates problems, and a high R-squared doesn't always mean your model is good.

A case in point is that the R-squared of 84% is high for this model, but the model is clearly inadequate because it doesn't describe the curvature.

When you have one explanatory variable, it's easy to see the curvature using a fitted line plot. However, with multiple regression, curved relationships are not always so apparent. For these cases, residual plots are a crucial indicator of whether your model adequately captures curved relationships.

In other cases, you might need to depend on subject-area knowledge to fit curves appropriately. Previous experience or research can tell you that the effect of one variable on another varies based on the value of the independent variable. Is there a limit, threshold, or point of diminishing returns where the relationship changes?

Let's examine a real example of a curved relationship and see how it affects our interpretation. I'll model body mass index (BMI) against percent body fat in 12 year old girls. These are actual data I gathered as part of a study.

We have only one independent variable (BMI), so we can use a fitted line plot to display its relationship with body fat percentage. The relationship between the variables is curvilinear. I'll use a polynomial term to fit the curvature. In this case, I'll include a quadratic (squared) term. The fitted line plot suggests that this model fits the data.

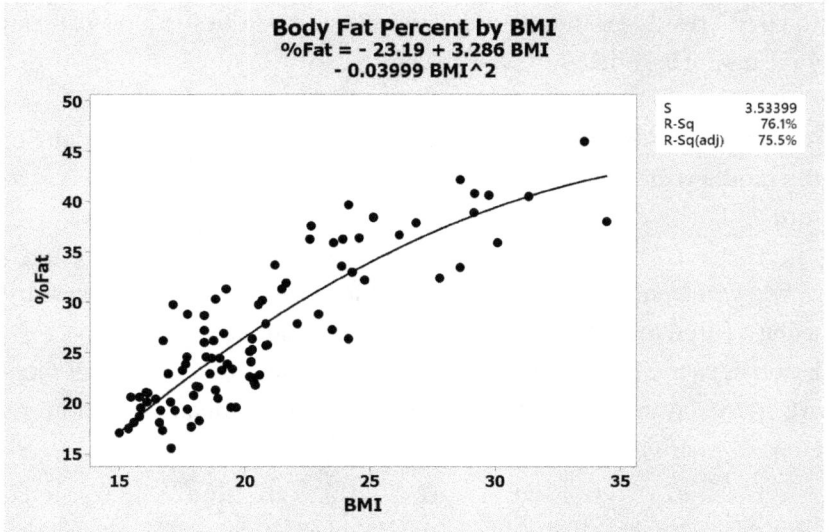

Body Fat Percent by BMI
%Fat = - 23.19 + 3.286 BMI
- 0.03999 BMI^2

S	3.53399
R-Sq	76.1%
R-Sq(adj)	75.5%

Notice the curved relationship. The fitted line follows the data nicely because the observations fall randomly around it for the entire range. The R-squared goodness-of-fit is 76.1%, which isn't fantastic but not terrible.

We tend to think of BMI scores in a linear sense. That is, if you start at either 16 or 30 and increase BMI by 1, we assume it represents the same increase in fat mass. The curved relationship shown above suggests that this is not true. The change in fat mass varies depending on the specific BMI value you start at.

At low BMIs, a single point increase from 15 to 16 relates to a larger increase in body fat percentage. You can see the curve is steeper in the low BMI region at the left side of the graph. However, the line flattens out as you move to the right. So, a single point increase from a BMI of 30 to 31 corresponds to a smaller body fat percentage change.

Regression analysis improves upon raw BMI scores for this population by correctly modeling the curvature.

As with confounding variables, curvature is something that analysts must adequately model to obtained unbiased results. However, it's possible to do that excessively where all you're doing is bending the regression line to fit the randomness in the data. The example below illustrates this problem.

These data come from the U.S. President ranking model in Chapter 7's discussion about handling outliers. I evaluated factors that correlate with the historians' ranking of U.S. Presidents. I was attempting (and succeeded!) at improving upon a model by Nate Silver. The full model contains more variables but here we'll look at the relationship between the highest approval rating that a U.S. President achieved and their rank by historians.

Imagine we are chasing a high R-squared and we fit the model using a cubic term that provides an S-shape. The coefficients all have statistically significant p-values too.

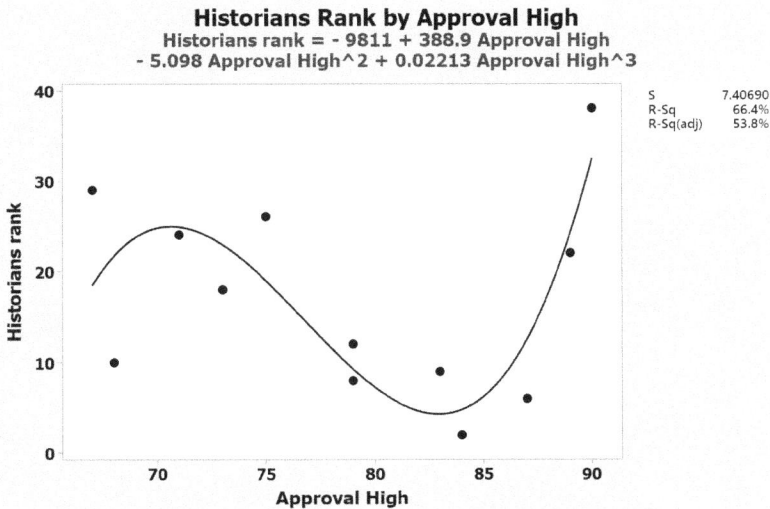

Historians Rank by Approval High
Historians rank = - 9811 + 388.9 Approval High
- 5.098 Approval High^2 + 0.02213 Approval High^3

S	7.40690
R-Sq	66.4%
R-Sq(adj)	53.8%

It sure looks like the line fits the data well, but it's fitting random noise. There is no plausible theoretical reason why the historian's rank should first decline with a rising approval rating, then increase for a while, only to descend again at the end. If that description appears to contradict the graph, recall that as historians' rank rises in value on the vertical axis, they're actually getting worse!

Again, theory should guide you. There is no theoretical basis for this pattern to exist in these data.

Additionally, while it goes beyond this book, the predicted R-squared for this model is zero. This specialized form of R-squared is a type of cross-validation that indicates the model will not fit additional data well even though it fits the current data. That's because it's fitting the random noise, essentially playing connect the dots! The later section about overfit models expands on this idea.

For my study, I found no correlation between these variables, as shown in the fitted line plot below. It's nearly a perfect example of no relationship because it is a flat line with an R-squared of 0.7%!

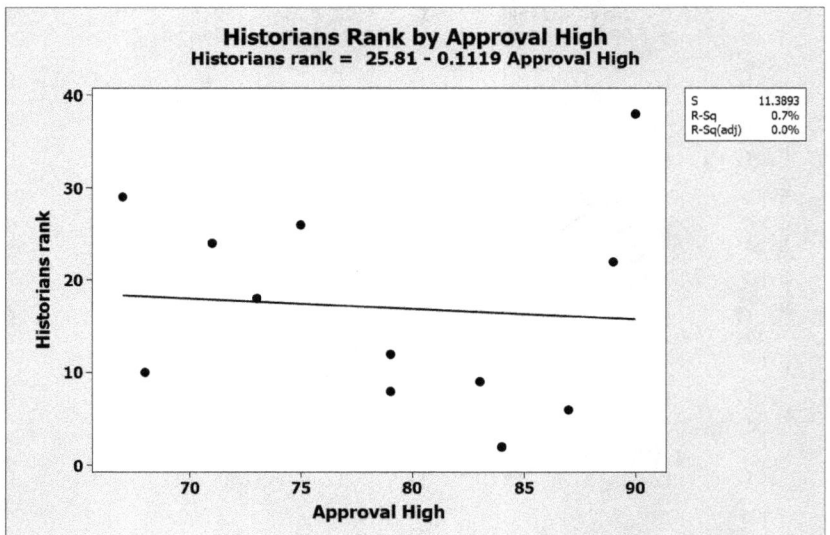

Historians Rank by Approval High
Historians rank = 25.81 - 0.1119 Approval High

It's incredibly tempting to go with the crazy bent line because it appears to explain most of the variability. I sure wanted to use it because it appears to be a better predictor than any in my final model. But it's random noise, not a signal. Hence, I didn't include the highest approval ratings in the model, but interestingly, each president's lowest approval rating was a significant predictor.

You can read my article here:
https://statisticsbyjim.com/regression/regression-model-presidential-rankings/

Interaction Effects

An interaction effect occurs when the effect of one variable depends on the value of another variable. Interaction effects are common in regression models, ANOVA, and designed experiments.

Many variables can affect the outcome of interest in a study, whether it's a taste test or a manufacturing process. Changing these variables can directly affect the outcome. For instance, changing the food condiment in a taste test can affect the overall enjoyment. In this manner, analysts use models to assess the relationship between each explanatory variable and the outcome. This kind of effect is called a main effect. While main effects are relatively straightforward, assessing only main effects can be a mistake.

In more complex study areas, the explanatory variables might interact with each other. Interaction effects indicate that a third variable influences the relationship between an independent and dependent variable. In this situation, statisticians say that these variables interact because the relationship between an explanatory and outcome variable changes depending on the value of a third variable. This type of effect makes the model more complex, but if the real world behaves this way, it is critical to incorporate it into your model. For example, the relationship between condiments and enjoyment probably depends on the type of food—as we'll see!

I think of interaction effects as an "it depends" effect. Let's use an intuitive example to help you conceptually understand these effects in an interaction model.

Imagine we are conducting a taste test to determine which food condiment produces the highest enjoyment. Our outcome is Enjoyment, and the two explanatory variables are both categorical: Food and Condiment.

Our model with the interaction term is:
Satisfaction = Food Condiment Food*Condiment

To simplify things, we'll include only two foods (ice cream and hot dogs) and two condiments (chocolate sauce and mustard) in our analysis.

Given the specifics of the example, an interaction effect would not be surprising. If someone asks you, "Do you prefer mustard or chocolate sauce on your food?" Undoubtedly, you will respond, "It depends on the type of food!" That's the "it depends" nature of an interaction effect. You cannot answer the question without knowing more about the other variable in the interaction term—which is the type of food in our example.

For this example, assume that the interaction between food and condiment is statistically significant. The best way to understand these effects is with a special type of line chart—an interaction plot. This type of plot displays the fitted values of the outcome variable on the y-axis, while the x-axis shows the values of an explanatory variable. Meanwhile, the various lines represent values of the second explanatory variable.

On an interaction plot, parallel lines indicate there is no interaction effect, while different slopes suggest one might be present. Below is the plot for Food*Condiment.

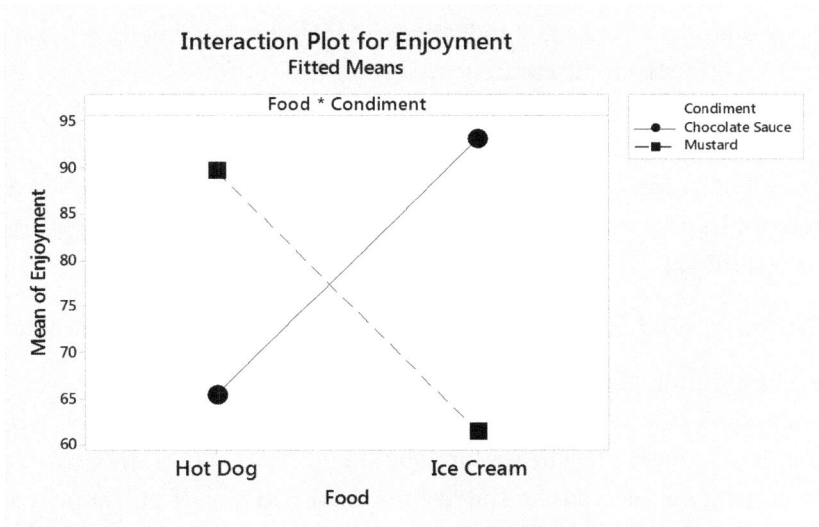

The crossed lines on the graph suggest that there is an interaction effect. The graph shows that enjoyment levels are higher for chocolate sauce when the food is ice cream. Conversely, satisfaction levels are higher for mustard when the food is a hot dog. You won't be happy if you put mustard on ice cream or chocolate sauce on hot dogs!

Which condiment is best? It depends on the type of food, and we've used statistics to demonstrate this effect.

When you have statistically significant interaction effects, you can't interpret the main effects without considering the interactions. In the previous example, you can only answer the question about which condiment is better by knowing the type of food. Again, "it depends."

Interaction effects can either enhance or moderate the influence of individual factors on an outcome. When factors interact to enhance an effect, they amplify each other's impact, creating a more substantial combined effect than either would have on its own. Conversely, when factors interact to moderate an effect, one factor can reduce or buffer the impact of the other, leading to a lesser combined effect than expected if the factors acted independently. Understanding these interactions is crucial for accurately predicting how different elements will influence results in real-world scenarios.

Let's look at some real interaction effects. In each case, notice how the relationship between the explanatory and outcome variable depends on a third variable.

Education and Socioeconomic Status in Preschool Education

A study conducted in Bahrain explored the interaction between socioeconomic status (SES) and preschool education on students' academic achievement. The researchers found that children from lower SES backgrounds who attended preschool had significant improvements in educational achievement, which were not as pronounced in children from higher SES backgrounds. This interaction suggests that preschool education particularly benefits children from disadvantaged socioeconomic backgrounds, helping to bridge the gap with their peers (Fateel, Mukallid, & Arora, 2021).

How much did preschool help the children? It wasn't a fixed amount across all groups. Instead, it *depends* on their SES.

High Protein Diet and Resistance Exercise on Fat-Free Mass During Weight Loss

A randomized controlled trial investigated the interaction between resistance exercise and different diets on preserving fat-free mass (FFM) during weight loss in overweight and obese older adults (Verreijen, Engberink, & Memelink, 2017).

The study revealed that resistance exercise combined with a high protein diet was more effective in preserving fat-free mass than exercising with other diets or either intervention alone. Specifically, the interaction between diet and exercise suggested that higher protein intake could enhance the effectiveness of resistance training in maintaining muscle mass during weight loss, providing a synergistic benefit that was not as pronounced when participants used either diet or exercise in isolation.

Again, it's an "it depends" effect. The relationship between resistance training and fat-free mass depends on the type of diet.

Interaction of Placement Characteristics and Emotional Experiences in Brand Recognition

A study examined how the characteristics of product placement in movies (like prominence, serial positions, and plot connection) and viewers' emotional experiences interact to affect brand recognition (Song, Chan, & Wu, 2019).

The findings indicated that both placement characteristics and emotional experiences independently influenced brand recognition. However, their interaction also played a significant role, showing that the context in which a brand is placed, combined with the scene's emotional impact, can significantly enhance or diminish brand recognition. These results suggest that marketers need to consider both the emotional response elicited by the placement and the placement's characteristics to optimize brand recognition strategies.

What is the relationship between product placement and brand recognition? It depends on the emotional context of the scene.

The idea of interactions probably makes sense to you. Particular combinations enhance or moderate the effect. Modeling them appropriately involves including the correct interaction terms. Suppose an interaction effect exists in the study area, but you don't include it in your model. In that case, the model is forced to attribute the interaction effects to the variables in the model.

That can be bad. Think back to our condiment preference model. If we didn't include the interaction effect, the model would be forced to determine that one condiment was best for both hot dogs and ice cream. Do you want mustard on your sundae or chocolate sauce on your hot dog?

Interaction effects are yet another case where the analyst must include them when the subject matter requires them to obtain unbiased results. Analysts can be tempted to include them when they're not needed but to artificially improve the statistical appearance of the results.

Overfitting and Chance Correlations

I've described regression as a seductive analysis because it is so tempting and so easy to add more variables, bend the regression line, and include interaction terms in the pursuit of improving how well the model fits the data. Every time you add a new variable or type of effect, the model *always* fits the data better. It's a mathematical certainty.

Problems occur when analysts explore a dataset and allow the patterns in the sample to suggest the correct model rather than being guided by theory. This process is easy because you can quickly test numerous combinations of explanatory variables to uncover statistically significant relationships. Automated model building procedures can fit thousands of models quickly. You can continue adding statistically significant variables as you find them, and the model's goodness-of-fit will always increase out of mathematical necessity, even if they don't reflect real effects or relationships.

Chance correlations exist, and excessive model fitting will find them and produce a model that appears to fit the data well but is full of false relationships and distorted results.

Overfitting a model is a different condition but produces the same problems as excessive model fitting. Overfitting occurs when a statistical model begins to describe the random error in the data rather than the relationships between variables. An overfit model has too many terms relative to the number of observations, causing it to fit the unique quirks of the specific sample rather than the broader patterns that might apply to other samples. This lack of generalizability means that while the model appears accurate for the analysts' data, it is unlikely to perform well on new observations.

Despite these problems, it can be hard to resist the allure of a high R-squared value. Understanding the issues that chasing a high R-squared and statistical significance can cause is crucial for avoiding them.

There are two fundamental problems with fitting excessing combinations of models and including variables that improve the fit:

- Following statistical significance opens the door to chance correlations.
- You include too many variables and types of effects.

Over the years, I've received numerous comments about how looking at many different variables, their interactions, and polynomials in all combinations makes sense. After all, what could possibly be wrong if you ended up with a model full of statistically significant variables that fit the data better?

Well, *everything* can be wrong!

These problems can produce misleading regression coefficients, p-values, and R-squared values.

I want these problems to sink in because they often occur when analysts chase a high R-squared and statistical significance. Indeed, inflated R-squared values are a *symptom* of overfit models! Despite the misleading results, it can be difficult for analysts to give up that nice high R-squared value.

That's why it's a form of p-hacking. You get deceptively good models with statistically significant p-values and a nice, high goodness-of-fit!

How severe is this problem?

I've written an article in which I generated random data for both the outcome and many explanatory variables. Then, I used standard, automated algorithms that mindlessly chased the goodness-of-fit for models that allegedly explain these randomly generated data. The resulting models look good with statistically significant variables, how well they fit the data, and satisfying the assumptions. There are no visible signs of problems, even though *all* the findings are deceptive.

Unfortunately, these results don't reflect actual relationships but instead represent chance correlations guaranteed to occur with enough opportunities. Opportunities that the automated procedure happily provided.

The regression model suggests that random data can explain other random data, which is impossible. If you didn't know that there are no actual relationships between these variables, these results would lead you to completely inaccurate conclusions.

You can read my article here:
https://statisticsbyjim.com/regression/data-mining-select-regression-models/

There are statistical tools that can help you navigate these issues, such as the predicted R-squared I mentioned earlier. However, these tools aren't perfect and can disagree with each other.

What are the solutions? They include several things I've already covered.

Sample size! Increase it sufficiently to handle whatever model complexity your subject area requires. That means you'll need subject-area knowledge in advance to have an informed idea.

Follow theory! Don't chase statistical significance and goodness-of-fit mindlessly. That also means you'll need subject-area knowledge.

The best statistical analyses always require using large amounts of subject-area knowledge.

AI Apex

As I write this book, the current state of artificial intelligence (AI) is nothing short of revolutionary. It's transforming how we live, work, and think about the possibilities of technology. In recent years, AI has achieved remarkable feats that once belonged to science fiction. Autonomous vehicles are navigating busy streets, AI systems are diagnosing diseases with accuracy comparable to or surpassing human experts, and virtual assistants are increasingly sophisticated, managing our schedules and even offering companionship.

Perhaps most impressively, AI has mastered complex games like Go, solving problems that require strategic thinking and intuition. These advancements are not just technical marvels but are paving the way for new industries and enhancing our ability to tackle some of the world's most pressing issues, from climate change to healthcare.

As AI continues to evolve, it promises to integrate further into our daily lives, reshaping society in unprecedented ways and offering exciting and scary prospects for future innovation. While AI's short- and long-run impacts on society are unknown, it's clear these tools are going to profoundly reshape our lives. I'm not going to make predictions, but I do want to address how AI fits into the Thinking Analytically pyramid.

At the current state of development, I'm placing AI in the Analytics level. AI tools build on statistical techniques and algorithms. The same issues that apply to all data analysis still apply to AIs. In the future, AIs might evolve so much that they require their own pyramid level.

Keep in mind that AI capabilities are growing exponentially, and we humans struggle to think exponentially. Remember doubling pennies for a month in Chapter 2? That's how AIs are growing. Be prepared for a mismatch between how quickly we expect their capacities to grow and their actual growth.

Let's look at how AIs relate to the Thinking Analytically pyramid.

Cognitive Foundation: AI can help mitigate human cognitive errors by providing consistent, objective analysis without the biases that typically affect human judgment. However, this help is similar to what other structured data analyses provide—*potential* objectivity. Achieving that objectivity depends on higher levels of the pyramid. Additionally, AI systems can inherit or amplify these biases if not carefully designed and monitored.

Probability Mindset: AI excels in environments where probabilities must be assessed and managed. Machine learning models, particularly those involving Bayesian networks or stochastic processes, can analyze vast datasets to predict outcomes more accurately than human intuition.

Data Quality: AI's performance heavily depends on the quality of data it processes. Poor data quality (e.g., incomplete, inaccurate, non-representative, or biased data) can lead to flawed AI outputs. Ensuring high-quality data is crucial for AI systems to function effectively and make reliable decisions.

Experimental Design: The design of experiments directly influences the conclusions AI can draw from data. AI trained on data from randomized experiments can more reliably infer causal relationships. In contrast, those trained on observational data will struggle with biases and confounding variables, limiting their ability to establish causality accurately.

Analytics: AI incorporates complex statistical algorithms to process and analyze data. Techniques such as deep learning involve layers of computation that can model intricate patterns not readily discernible by traditional statistical methods. However, understanding these models' inner workings and ensuring they correctly interpret the data is essential for reliable results.

As you can see, AIs can be a valuable tool by shoring up some of our natural shortcomings in the lower two levels of the pyramid. However, it is susceptible to the same issues as other forms of data analysis in the upper three levels. Garbage in, garbage out applies to AIs as much as to other forms of analysis.

Artificial intelligence represents the cutting edge in leveraging data for decision-making. However, it is sensitive to the challenges I discuss throughout this book. Despite their computational prowess, the quality and nature of their input data fundamentally constrains AI systems. Issues such as non-representative sampling, biased data inputs, and overlooked confounders can severely undermine the effectiveness and fairness of AI-driven solutions.

AI also requires careful oversight to ensure it adheres to rigorous analytical standards because it tends to hallucinate answers. Many of these AI tools are "black boxes" from which we don't understand how they derive the answers.

Let's examine several examples of AI gone wrong. Notice how they all involve some form of data bias that produces biased results.

Amazon Hiring Tool

This AI was developed to streamline recruitment by evaluating resumes. However, because it was trained primarily on historical data that reflected a male-dominated tech industry, it inadvertently learned to favor male candidates. The system downgraded resumes mentioning "women's chess club captain" or containing other female-associated terms. This bias reinforced gender disparities, counteracting efforts to promote diversity and inclusion within the company.

CheXNet

Stanford's AI for pneumonia detection from chest X-rays exemplifies the challenges of non-representative sampling. Originally trained on a dataset from specific populations, it struggled to accurately diagnose pneumonia in patient groups from different demographic and geographic backgrounds. This failure to generalize well across diverse populations resulted in potential misdiagnoses.

PredPol

This predictive policing tool used historical crime data to forecast crime hotspots. However, because the underlying data often reflected a disproportionate focus on minority neighborhoods (i.e., over-policing), the tool perpetuated these biases. This led to increased police patrols and surveillance in these areas, exacerbating community tensions and trust issues between law enforcement and minority communities and potentially diverting resources from other areas in need.

These examples underscore the necessity for scrutiny in developing and deploying AI systems. Ensuring that AI solutions are fair, equitable, and truly reflective of the diverse populations they serve is not just a technical challenge but a profound ethical imperative in the modern era of data-driven decision-making.

Closing Thoughts

I hope this short foray into modeling outcomes in a multivariate environment was informative. Controlling confounders, modeling curvature and interaction effects, and letting theory guide your modeling rather than algorithms are all crucial for creating useful models. Hopefully, you got a taste of what is involved and some of the inherent temptations and problems. As an analytical thinker, these are issues you must consider.

"All models are wrong, but some are useful." –George E. P. Box

The flexibility of model fitting is a double-edged sword. Analysts can fit models to a wide array of data and include the variables, curvature, and interaction effects as the subject matter requires. Unfortunately, analysts can abuse that flexibility to bend the results to a desired outcome, either unintentionally or intentionally.

As we worked our way through the thinking analytically pyramid, it was fascinating to see how the cognitive biases on the base layer not only prompted statisticians to develop various statistical methods to overcome those limitations but also how those cognitive errors find the freedom to manifest in the flexibility of multivariate modeling.

That's why statisticians recommend developing a fixed plan before collecting and analyzing your data. It helps you avoid the temptation to tinker with the analysis to get the desired results.

Throughout this book, we've examined the various challenges and limitations of analytical thinking as we climbed the thinking analytically pyramid. Let's recap the pyramid layers briefly.

1. **Cognitive Foundation**: Human brains have common cognitive errors that distort our attempts to see the world objectively.

2. **Probability Mindset:** We're not good at intuitively assessing probabilities to make objective decisions.
3. **Data Quality:** How you collect and measure your data affects the results. Do you trust your data?
4. **Experimental Design:** Experimental designs determine which conclusions analysts can draw and how confident we can be that outside variables aren't distorting the results.
5. **Analytics:** The nitty-gritty details can and do affect the results.

Unfortunately, our brains have limitations built into them. We have a difficult time recalling and weighing all data appropriately. Frequently, emotions and narrative details greatly distort this process. I covered these cognitive biases and problems with calculating probabilities early in this book.

Statisticians devised methods for overcoming these innate shortcomings. These include sampling methods, a framework for evaluating the reliability and validity of measurements, experimental designs, and various statistical assessments.

Being vigilant about biases and methodological shortcomings is crucial for drawing valid conclusions from data. As discussed in earlier chapters, biases can significantly skew our understanding and interpretation of data. Confirmation and selection bias, among many others, can distort our view of reality. That's true even in big data contexts.

Confirmation bias occurs when individuals favor information or interpretations that confirm their preexisting beliefs or hypotheses. It can lead to selective data gathering, overemphasizing supporting evidence, and disregarding contradictory information. The impact is particularly profound in research settings where the desire to find a specific outcome can lead to unintentional cherry-picking of data, potentially leading to false conclusions and misleading claims.

Selection bias arises when the data collection process ensures that the sample is not representative of the population intended to be analyzed. This problem occurs with the non-random selection of participants or when participants with specific characteristics drop out disproportionately. The result is that the findings might not be generalizable to the broader population, thus limiting the scope and applicability of the analysis.

On the pyramid's Experimental Design level, you saw how the study's context affects the conclusions you can draw. A lack of proper control groups or failure to account for confounding variables can lead to erroneous causality attributions. For instance, without randomization in experimental design, observed effects might be attributable to preexisting differences rather than the intervention or treatment studied.

Then, you climb to the statistical level of the pyramid, where you can get the wrong answers for things like using improper methods for seemingly mundane aspects (outliers, missing data, etc.), variability, random chance, and incorrect modeling. Shoot, even just fitting too many models is potentially problematic. Many things can go wrong at all stages, causing you to draw erroneous conclusions.

Data mismanagement, whether through incorrect data entry, inappropriate data cleaning practices, or mishandling of missing data and outliers, can lead to incorrect analyses and faulty conclusions. Analysts should document all data handling processes to ensure the integrity of the research findings.

Small sample sizes can produce unreliable and unstable estimates that inaccurately reflect the larger population. This problem makes findings susceptible to substantial variability, potentially rendering them non-replicable in larger, more representative samples. Unfortunately, our minds are wired to remember these surprising and extreme results more strongly than typical ones.

Real-world variability is a form of uncertainty that affects our ability to draw conclusions. Even when analysts do everything correctly, random chance can garble our results.

Awareness of these biases and methodological pitfalls is the first step in safeguarding the accuracy of your analyses. This book explored these issues in detail, giving you the mindset to critically assess and enhance the robustness of your own and others' analytical conclusions.

As you think analytically and engage with data, continually question the processes and assumptions underlying the analyses and seek to correct and refine your approach where necessary. This vigilance enables you to produce more reliable and credible results, strengthening the overall impact of your research.

Unfortunately, there are obstacles at every turn. If you want to use the best sampling method to obtain a truly random sample, it will be more costly and time-consuming. Obtaining the best measurements is similarly more expensive. The most effective experimental designs involve elaborate environments and procedures to minimize outside sources of variation and biases. To establish causal relationships, you'll need to use random assignment, which isn't always possible.

In short, it's difficult to understand the complexities of the real world using relatively small samples. Big data solves the sample size problem but not the other aspects that introduce bias. Sometimes, the best you can do is recognize the limitations and appropriately adjust your understanding of the analytical results. This awareness helps prevent these shortcomings from misleading you.

After reading this book and all the pitfalls I cover, you probably aren't surprised by the replication crisis I mentioned in the p-hacking section. The same goes for the infamous article *Why Most Published Research Findings Are False* by John P. A. Ioannidis.

But this book isn't just about how things can go wrong. It provides a framework for thinking analytically and evaluating information, whether receiving information in the wild, reading about other analyses, or performing it yourself. Or perhaps someone is trying to influence you with the latest study?

Watch out for your biases, especially for findings that agree with your existing beliefs. That's where some of your blind spots likely live! Beware of biases in other analysts and for potential methodological shortcomings.

Despite all the difficulties, I believe in science as a self-correcting method. Scientific analyses are published, scrutinized, and subsequently refined or corrected as necessary. Falsified results won't systematically replicate for other researchers.

It's not a linear process because even the most brilliant scientists can make mistakes. However, follow-up research will identify problems and propose solutions. I've shown you multiple examples of how preliminary research has gotten it wrong, but later research used better methods and analyses to find the correct answers.

Think back to the changing coffee findings in Chapter 5. At first, scientists thought coffee was generally bad for your health. They reported the findings and discussed their methods. Other scientists read those articles and suggested better analyses that accounted for confounders. They published new results suggesting that coffee is good for your health. And now that's the consensus after some discussion.

Science is a self-correcting field that encourages transparency, scrutiny, discussion, and change. Indeed, scientists themselves identified and reported the replication and false findings crises. And now there's a movement to address these issues.

No single study can prove anything, and no single analyst can get everything correct. Instead, you need multiple experiments involving teams of researchers. At its best, science is a communal effort where you have groups of highly informed experts around the world performing experiments, critiquing the work of their peers, finding better methods, and attempting to replicate results.

I'm more concerned about big data in business and government. Analysts in those areas tend not to publicize their methodology and findings as broadly as scientists.

In big data, it's crucial to recognize the distinct challenges posed by its application within business and government sectors. Unlike scientific research, where expert peers scrutinize methodologies and findings, the practices in these non-scientific contexts often remain opaque. This lack of transparency can obscure significant biases and methodological errors, potentially leading to decisions that adversely affect millions of lives.

In the business world, companies leverage big data for everything from marketing strategies to operational efficiencies to health insurance decisions. However, without a clear view of the algorithms and data sets used, consumers and regulators are left in the dark about how decisions that influence consumer behavior and privacy are made. Similarly, in government, big data can inform policies on everything from healthcare to criminal justice. Yet, if these data-driven policies are not open to public review, they risk perpetuating existing inequalities and injustices under the guise of impartiality.

The need for accountability is paramount. Governments must adopt standards that require the disclosure of methodologies and data sources used in their analyses. Such standards will foster trust and enable independent verification of the results, ensuring that decisions are fair and just.

By integrating these principles, businesses and governments can mitigate the risks of bias and ensure that big data tools enhance rather than compromise societal welfare.

For analysts deeply embedded in big data, recognize your work's profound impact, whether it's driving business strategies or informing governmental policies. The responsibility that accompanies the handling of such data is immense, given its potential to influence public behavior and crucial policy decisions.

In recognizing these challenges, your role goes beyond mere data analysis. Continuously question and critique your methods and data integrity, as detailed throughout this book. By understanding and addressing the pitfalls, you better position yourself to produce trustworthy results.

For individuals evaluating analyses and arguments, slow down! Take the time to get a broad perspective to guard against cognitive biases. Be sure to fully understand all the details and complexities of the subject matter. After reading this book, it should be clear that all the details are crucial, and they can flip your interpretation. Use the full scope of your thinking analytically mindset. Don't make snap judgments—particularly when they reinforce what you already believe.

As we conclude this exploration of the intricacies and pitfalls of thinking analytically, remember that the power to discern, critique, and ultimately understand data lies within your grasp. This book has equipped you with the knowledge to navigate the vast seas of information, enabling you to recognize biases, and to identify common methodological errors that can distort the truth. Armed with these insights, you are not just a passive consumer of data but an active participant in the pursuit of knowledge.

Every chapter has laid a foundation for skepticism—not cynicism, but of inquiry and diligence. As you apply these principles, whether evaluating a new study, scrutinizing a report, or conducting your analyses, you wield the tools necessary to cut through the noise and uncover deeper truths. Let this knowledge liberate you from the constraints of unexamined data and guide you toward more informed decisions and opinions.

Embrace this empowerment as your intellectual toolkit expands, and approach each new piece of information with a critical yet open mind. This mindset is not just about avoiding errors or countering misleading statistics. It's about enhancing your understanding of the world.

Carry the torch of informed skepticism and let it illuminate your path in your professional endeavors and everyday life. Remember, in a world awash with data, a well-honed analytical mind is your best defense against the tide of misinformation.

My Other Books

Introduction to Statistics: An Intuitive Guide

Learn statistics without fear! Build a solid foundation in data analysis. Be confident that you understand what your data are telling you and that you can explain the results to others! I'll help you intuitively understand statistics by using simple language and deemphasizing formulas.

This guide starts with an overview of statistics and why it is so important. We proceed to essential statistical skills and knowledge about different types of data, relationships, and distributions. Then we move to using inferential statistics to expand human knowledge, how it fits into the scientific method, and how to design and critique experiments—whether it's your own or another researcher's.

Learn the fundamentals of statistics in this 255-page book:

- Why is the field of statistics vital in our data-driven society?
- Interpret graphs and summary statistics.
- Find relationships between different types of variables.
- Understand the properties of data distributions.
- Use measures of central tendency and variability.
- Interpret correlations and percentiles.
- Use probability distributions to calculate probabilities.
- Learn about the normal and binomial distributions in depth.
- Grasp the differences between descriptive and inferential statistics.
- Use data collection methodologies properly and understand sample size considerations.
- Access free downloadable datasets so you can try it yourself.

Currently available in print and as an ebook at most retailers!
Learn more on my website: statisticsbyjim.com/store

Hypothesis Testing: An Intuitive Guide

Build a solid foundation for understanding how hypothesis tests work and become confident that you know when to use each type of test, how to use them properly to obtain reliable results, and interpret the results correctly. Chances are high that you'll need a working knowledge of hypothesis testing to produce new findings yourself and to understand the work of others. I present a wide variety of tests that assess characteristics of different data types. I focus on helping you grasp key concepts, methodologies, and procedures while deemphasizing equations. Learn how to use these tests painlessly!

In today's data-driven world, we hear about making decisions based on the data all the time. Hypothesis testing plays a crucial role in that process, whether you're in academia, making business decisions, or in quality improvement. Without hypothesis tests, you risk drawing the wrong conclusions and making bad decisions. The world today produces more data and more analyses designed to influence you than ever before. Are you ready for it?

In this 367-page book, build the skills and knowledge you'll need for effective hypothesis testing, including the following:

- Why you need hypothesis tests and how they work.
- Using significance levels, p-values, confidence intervals.
- Interpreting the results.
- Select the correct type of hypothesis test to answer your question.
- Learn how to test means, medians, variances, proportions, distributions, counts, correlations for continuous and categorical data, and outliers.
- One-Way ANOVA, Two-Way ANOVA and interaction effects.
- Estimate a good sample size for your study.
- Checking assumptions and obtaining reliable results.

- Manage the error rates for false positives and false negatives.
- Understand sampling distributions, central limit theorem, and statistical power.
- Know how t-tests, F-tests, chi-squared tests, and post hoc tests work.
- Learn about the differences between parametric, nonparametric, and bootstrapping methods.
- Examples of different types of hypothesis tests.
- Downloadable datasets so you can try it yourself.

Currently available in print and as an ebook at most retailers!
Learn more on my website: statisticsbyjim.com/store

Regression Analysis: An Intuitive Guide

Over the course of this full-length book, you'll progress from a beginner to a skilled practitioner. I'll help you intuitively understand regression analysis by focusing on concepts and graphs rather than equations and formulas. I use everyday language so you can grasp regression at a deeper level.

Learn practical tips for performing your analysis and interpreting the results. Feel confident that you're analyzing your data properly and able to trust your results. Know that you can detect and correct problems that arise.

This 336-page book covers the following:

- How regression works and when to use it.
- Selecting the correct type of regression analysis.
- Specifying the best model.
- Understanding main effects, interaction effects, and modeling curvature.
- Interpreting the results.
- Assessing the fit of the model.
- Generating predictions and evaluating their precision.
- Checking the assumptions and resolving issues.
- Examples of different types of regression analyses.
- Downloadable datasets for the examples.

Currently available in print and as an ebook at most retailers!
Learn more on my website: statisticsbyjim.com/store

References

Caputo, A. (2014). Relevant information, personality traits and anchoring effect. *International Journal of Management and Decision Making*, 62-76.

Croson, R., & Sundali, J. (2005). The gambler's fallacy and the hot hand: Empirical data from casinos. *Journal of Risk and Uncertainty*, 195-209.

Dietrich, D., & Olson, M. (1993). A Demonstration of Hindsight Bias Using the Thomas Confirmation Vote. *Psychological Reports*, 377-378.

Englich, B., & Soder, K. (2009). Moody experts--How mood and expertise influence judgmental anchoring. *Judgment and Decision Making*, 41.

Fateel, M. J., Mukallid, S., & Arora, B. (2021). The Interaction Between Socioeconomic Status and Preschool Education on Academic Achievement of Elementary School Students. *International Education Studies*, 60-66.

Fischoff, B. (1975). Hindsight ≠ Foresight: The Effect of Outcome Knowledge on Judgment Under Uncertainty. *Journal of Experimental Psychology: Human Perception and Performance*, 288-299.

Herndon, T., Ash, M., & Pollin, R. (2014). Does High Public Debt Consistently Stifle Economic Growth? A Critique of Reinhart and Rogoff. *Cambridge Journal of Economics*, 257-279.

Hughes, K., Thompson, J., & Trimble, J. E. (2016). Investigating the Framing Effect in Social and Behavioral Research: Potential Influences on Behavior, Cognition and Emotion. *Social Behavior Research and Practice*, 34-37.

Ioannidis, J. P. (2005). Why Most Published Research Findings Are False. *PLOS Medicine*, https://doi.org/10.1371/journal.pmed.0020124.

Jay, K. L., & Jay, T. B. (2015). Taboo word fluency and knowledge of slurs and general perjoratives: Deconstructing the poverty-of-vocabulary myth. *Language Sciences*, 251-259.

Kahneman, D. (2011). *Thinking, Fast and Slow*. Straus and Girouz.

Kruger, J., & Dunning, D. (1999). Unskilled and unaware of it: How difficulties in recognizing one's own incompetence lead to inflated self-assessments. *Journal of Personality and Social Psychology*, 1121-1134.

MacMahon, B., Yen, S., Trichopoulos, D., Warren, K., & Nardi, G. (1981). Coffee and cancer of the pancreas. *N Engl J Med*, 630-3.

Mursu, J. R. (2011). Dietary Supplements and Mortality Rate in Older Women: The Iowa Women's Health Study. *Arch Intern Med.*, 1625-1633.

Open Science Collaboration. (2015). Estimating the reproducibility of psychological science. *Science*.

Reinhart, C. M., & Rogoff, K. S. (2010). Growth in a Time of Debt. *American Economic Review*, 573-578.

Simonsohn, U., Nelson, L. D., & Simmons, J. P. (2013). P-Curve: A Key to the File Drawer. *Journal of Experimental Psychology: General*, 534-547.

Song, S., Chan, F. F., & Wu, Y. (2019). The interaction effect of placement characteristics and emotional experiences on consumers' brand recognition. *Asia Pacific Journal of Marketing and Logistics*, 1269-1285.

Stefan, A. M., & Schönbrodt, F. D. (2023). Big little lies: A compendium and simulation of p-hacking strategies. *Royal Society Open Science*, https://doi.org/10.1098/rsos.220346.

Tjønneland, A., Grønbaek, M., Stripp, C., & Overvad, O. (1999). The connection between food and alcohol intake habits among 48,763 Danish men and women. A cross-sectional study in the project "Food, cancer and health". *Ugeskrift for Laeger*, 6923-6927.

Tversky, A., & Kahneman, D. (1973). Availability: A heuristic for judging frequency and probability. *Cognitive Psychology*, 207-232.

Tversky, A., & Kahneman, D. (1973). On the psychology of prediction. *Psychological Review*, 237-257.

Tversky, A., & Kahneman, D. (1974). Judgment under uncertainty: Heuristics and biases. *Science*, 1124-1131.

Tversky, A., & Kahneman, D. (1981). The framing of decisions and the psychology of choice. *Science*, 453-458.

Tversky, A., & Kahneman, D. (1983). Extension versus intuitive reasoning: The conjunction fallacy in probability judgment. *Psychological Review*, 293-315.

Verreijen, A., Engberink, M., & Memelink, R. (2017). Effect of a high protein diet and/or resistance exercise on the preservation of fat free mass during weight loss in overweight and obese older adults: a randomized controlled trial. *Nutrition*, https://doi.org/10.1186/s12937-017-0229-6.

Wang, G., & Wu, L. (2020). Healthy People 2020: Social Determinants of Cigarette Smoking and Electronic Cigarette Smoking among Youth in the United States 2010 - 2018. *International Journal of Environmental Research and Public Health*.

Welsh, M. B., Delfabbro, P. H., Burns, N. R., & Begg, S. H. (2014). Individual differences in anchoring: Traits and experience. *Learning and Individual Differences*, 131-140.

Wertheimer, N., & Leeper, E. (1979). Electrical wiring configurations and childhood cancer. *Am J Epidemiol*, 273-84.

Yousaf, Z., Siddiqui, M. Y., Mushtaq, K., Feroz, S. E., Kammar, S. A., Mohamedali, M. G., & Chaudhary, H. (2020). Avoiding anchoring bias in the times of the pandemic. *Case Reports in Neurology*, 359-364.

Recommended Citation for This Book

Frost, J. (2024). Thinking Analytically: A Guide for Making Data-Driven Decisions. Statistics By Jim Publishing.

Index

About the Author

I'm Jim Frost, and I have extensive experience in academic research and consulting projects. In addition to my statistics website, I am a regular columnist for the American Society of Quality's *Statistics Digest*. Additionally, my most recent journal publication as a coauthor is *The Neutral Gas Properties of Extremely Isolated Early-Type Galaxies III* (2019) for the American Astronomical Society.

I've been the "data/stat guy" for research projects that range from osteoporosis prevention to analysis of online user behavior. My role has been to design the proper research settings, collect a large amount of valid measurements, and figure out what it all means. Typically, I'm the first person on the project to learn about new findings while interpreting the results of the statistical analysis. Even if the findings are not newsworthy, that thrill of discovery is an awesome job perk!

I love statistics and analyzing data! I've been performing statistical analysis on-the-job for 20 years and helping people learn statistics for over ten years at a statistical software company. I love talking and writing about statistics. While working at the statistical software company, I learned how to present statistics in a manner that makes statistics more intuitive.

I want to help you learn statistics. But I'm not talking about learning all the equations. Don't get me wrong. Equations are necessary. Equations are the framework that makes the magic, but the truly fascinating aspects are what it all means. I want you to learn the true essence of statistics. I'll help you intuitively understand statistics by focusing on concepts and graphs. After all, you use statistical software so you don't have to worry about the formulas and instead focus on understanding the results.

I've spent over a decade working at a major statistical software company. When you work on research projects, you generally use a regular group of statistical analyses. However, when you work at a statistical software company, you need to know of all the analyses that are in the software! I helped people use our software to gain insights and maximize the value of their own data regardless of their field.

Statistics is the field of learning from data. That's amazing. It gets to the very essence of discovery. Statistics facilitates the creation of new knowledge. Bit by bit, we push back the frontier of what is known. That is what I want to teach you! My goal is to help you to see statistics through my eyes—as a key that can unlock discoveries that are in your data.

The best thing about being a statistician is that you get to play in everyone's backyard. —John Tukey

I enthusiastically agree! If you have an inquisitive mind, statistical knowledge, and data, the potential is boundless. You can play in a broad range of intriguing backyards!

That interface between a muddled reality and obtaining orderly, valid data is an exciting place. This place ties together the lofty goals of scientists to the nitty-gritty nature of the real world. It's an interaction that I've written about extensively in this book and on my blog, and I plan to continue to do so. It's where the rubber meets the road.

One of the coolest things about the statistical analysis is that it provides you with a toolkit for exploring the unknown. Christopher Columbus needed many tools to navigate to the New World and make his discoveries. Statistics are the equivalent tools for the scientific explorer because they help you navigate the sea of data that you collect.

You'll be increasingly thankful for these tools when you see a worksheet filled with numbers and you're responsible for telling everyone what it all means.

Read more on my website: StatisticsByJim.com